W0036666

# Ethaphon: Impact on Sugarcane Physiology& Sugar Production

# Ethaphon: Impact on Sugarcane Physiology& Sugar Production

Editors

**VD Shinde, Ravindra Borade**

# Ethaphon: Impact on Sugarcane Physiology& Sugar Production

Edited by **VD Shinde, Ravindra Borade**

Printed in 2017

ISBN: 978-1-68117-061-9

Library of Congress Control Number: 2015935459

© 2016 by
SCITUS Academics LLC,
616, Corporate Way, Suite 2, 4766,
Valley Cottage, NY 10989

www.scitusacademics.com

This book contains information obtained from highly regarded resources. Copyright for individual articles remains with the authors as indicated. All chapters are distributed under the terms of the Creative Commons Attribution License, which permits unrestricted use, distribution, and reproduction in any medium, provided the original author and source are credited.

**Notice**

Reasonable efforts have been made to publish reliable data and views articulated in the chapters are those of the individual contributors, and not necessarily those of the editors or publishers. Editors or publishers are not responsible for the accuracy of the information in the published chapters or consequences of their use. The publisher believes no responsibility for any damage or grievance to the persons or property arising out of the use of any materials, instructions, methods or thoughts in the book. The editors and the publisher have attempted to trace the copyright holders of all material reproduced in this publication and apologize to copyright holders if permission has not been obtained. If any copyright holder has not been acknowledged, please write to us so we may rectify.

# Contents

# Preface

Ethephon is the most widely used plant growth regulator. Upon metabolism by the plant, it is converted into ethylene, a potent regulator of plant growth and ripeness. It is often used on wheat, coffee, tobacco, cotton, and rice in order to help the plant's fruit reach ripeness more quickly. Cotton is the most important single crop use for ethephon. It initiates fruiting over a period of several weeks, promotes early concentrated boll opening, and enhances defoliation to facilitate and improve efficiency of scheduled harvesting. Harvested cotton quality is improved. Ethephon also is widely used by pineapple growers to initiate reproductive development force of pineapple. Ethephon is also sprayed on mature-green pineapple fruits to degreen them to meet produce marketing requirements. There can be some detrimental effect on fruit quality. Sugarcane is a worldwide crop cultivated in 105 countries. From its very origin in earlier times to its present day production sugarcane has played its role in improving socioeconomic conditions of human society. It is mainly used as a food crop for the production of raw and refined sugar, gur, shakkar and molasses. Improved cane varieties and production practices have been optimized for increase of cane yield per hectare in the fields and maximizing sugar recoveries in the sugar factories. Sugarcane has gained importance for its dietary value and its industrial utilization for a number of products. The crop is of immense economic importance for the prosperity of people. Its products and bi-products have revolutionized the native and international trade and the crop production trends have played dominant role in altering the economic and fiscal position of countries.

**Editor**

Chapter 1

# Efficient Growth of kluyveromyces Marxianus Biomass used as a Biocatalyst in the Sustainable Production of Ethyl Acetate

Christian Löser[1], Thanet Urit[1,2], Erik Gruner[1], and
Thomas Bley[1]

[1]Institute of Food Technology and Bioprocess Engineering, TU Dresden,
Dresden, 01062, Germany

[2]Department of Biology and Biotechnology, Faculty of Science and
Technology, Nakhon Sawan Rajabhat University, Nakhon Sawan,
60000, Thailand

# ABSTRACT

## Background

Whey is just turning from a waste of milk processing to a renewable raw material in biotechnology for producing single-cell protein, bioethanol, or ethyl acetate as an economic alternative. Conversion of whey-borne sugar into ethyl acetate requires yeast biomass as a biocatalyst. A high cell concentration results in a quick ester synthesis, but biomass growth means consumption of sugar at the expense of ester production. Efficient and cost-saving biomass production is thus a practical requirement. Whey is poor in nitrogen and has therefore to be supplemented with a bioavailable N source.

## Methods

Several aerobic growth tests were performed with *Kluyveromyces marxianus* DSM 5422 as a potent producer of ethyl acetate in whey-borne media supplemented with various N sources. Preliminary tests were done in shake flasks while detailed studies were performed in a stirred bioreactor.

## Results

Ammonium sulfate resulted in strong acidification due to remaining sulfate, but costly pH control increases the salt load, being inhibitory to yeasts and causing environmental impacts. Ammonium carbonate lessened acidification, but its supplement increased the initial pH to 7.5 and delayed growth. Urea as an alternative N source was easily assimilated by the studied yeast and avoided strong acidification (much less base was required for pH control). Urea was assimilated intracellularly rather than hydrolyzed extracellularly by urease. Conversion of urea to ammonium and usage of formed ammonium for biomass production occurred with a similar rate so that the amount of excreted ammonium was small. Ammonium hydroxide as another N source was successfully added by the pH controller during the growth of *K. marxianus* DSM 5422, but the medium had to be supplemented with some ammonium sulfate to avoid sulfur limitation and to initiate acidification. Non-limited growth resulted in 82 mg N per g of biomass, but N-limited growth diminished the N content.

## Conclusions

*K. marxianus* could be efficiently produced by supplementing the whey with nitrogen. Urea and ammonia were the favored N sources due to the proton neutrality at assimilation which lessened the salt load and reduced the supply of alkali for pH control or made this even needless.

# BACKGROUND

Sustainable and environmentally compatible development requires successive substitution of fossil resources by renewable raw materials. This applies to the energy sector but is also true for production of industrial bulk materials. Such a bulk chemical is ethyl acetate with an annual world production of 1.7 million tons [1]. Ethyl acetate is an organic solvent of moderate polarity with versatile industrial applications. Another prospective application is the biodiesel production from vegetable oil; here, triglycerides are transformed to fatty acid ethyl esters in a lipase-catalyzed transesterification reaction with ethyl acetate as an acyl acceptor instead of methanol [2]-[5]. Although being an irritant and intoxicant at higher concentrations, ethyl acetate is less toxic to humans compared to many other solvents. Ethyl acetate is an environmentally friendly compound since the ester is easily degraded by bacteria [6]-[8] and is regarded as a non-persistent pollutant of the atmosphere [9].

Synthesis of ethyl acetate currently proceeds by petrochemical processes, which are based on crude oil constituents or natural gas, run at elevated temperature and pressure and commonly require catalysts [9]. The reactions are often incomplete, and the recovery of ethyl acetate and residual precursors needs a high input of energy.

Microbial production of ethyl acetate from renewables could become an interesting alternative. Various yeast species can synthesize ethyl acetate (reviewed by Löser et al. [9]), but only *Pichia anomala*, *Candida utilis*, and *Kluyveromyces marxianus* produce this ester in larger amounts. *K. marxianus* is the most promising candidate for large-scale ester production since this dairy yeast with GRAS status grows quickly, converts sugar directly into ethyl acetate without ethanol as an essential intermediate, and produces the ester with a high rate and yield [9]-[17]. The ester synthesis in *K. marxianus* is easy to control by the level of iron [11],[13],[16]-[18]. *K. marxianus*exhibits a distinct thermal tolerance which allows cultivation at an elevated temperature [14],[19],[20] which in turn accelerates the ester stripping and advances its process-integrated recovery. The outstanding capability of *K. marxianus* for lactose utilization offers the chance for using whey as a resource of ester synthesis. *K. marxianus* converts whey-borne lactose with a high yield into ethyl acetate [9], but its ability for sucrose and glucose assimilation [19]-[21] enables production of ethyl acetate with this yeast from renewables like sugarcane, grain, and corn.

Ethanol is another product of microbial sugar conversion, but several factors favor microbial ester over ethanol production [22]: the higher market price of the ester, a reduced number of process stages, a faster process, and a cost-saving product recovery.

Conversion of sugar into ethyl acetate requires yeast biomass as a biocatalyst. A high biomass concentration results in a quick process, but the production of this biomass, on the other hand, is connected with sugar consumption which reduces the portion of sugar available for ester production (as demonstrated in pilot-scale experiments [13]). The right balance between yeast growth and ester synthesis or, in other words, a compromise between a quick process and a high ester yield is of practical importance. This balance can be controlled by the available iron [13],[16],[23].

An efficient production of the required yeast biomass is an important factor for the economy of the total process of ester and ethanol production. At whey-based bio-ethanol production, the biomass is often considered as a gift and sugar utilization for biomass production is ignored. Such an 'out-sourcing' of biomass processing only seemingly improves the economy.

Much research was done in the field of whey-based *K. marxianus* cultivation for single-cell protein production [24]-[28]. The results refer to some problems at cultivation of *K. marxianus* in whey-based media; whey is poor in bioavailable nitrogen so that nitrogen can limit yeast growth [26]-[29]. Nitrogen-limited growth can even deregulate yeast metabolism and provoke ethanol formation at aerobic conditions [30]. Whey was often supplemented with ammonium as a source of nitrogen to stimulate growth of *K. marxianus*[10]-[14],[16],[27],[29],[31]-[37]. Added ammonium increased the yield of biomass[27],[32],[35] or was without significant effect [29],[31],[36]. Ammonium is usually supplemented in the form of ammonium sulfate where ammonium is intensively consumed while most of the sulfate remains in the medium and causes an ionic imbalance and acidification [15],[38]-[41]. Such acidification can be inhibitory to *K. marxianus* since its growth rate is distinctly reduced at pH $\leq$3.5 [22],[28],[42],[43]. Supplementation of whey with $(NH_4)_2SO_4$ or $NH_4Cl$ requires cost-intensive pH correction with NaOH or KOH which in turn increases the salt load and thus inhibits yeast growth [31],[42] and creates waste-water problems.

Urea is an alternative N source since growth with urea exhibits proton neutrality without significant pH changes [38]-[40]. Assimilation of urea by yeasts is not an exception but the rule; 122 of 123 tested yeasts were able to metabolize urea [44]. There are several potential advantages of urea[25],[28],[39]: a high amount of nitrogen per unit weight, a low price, and a reduced or even omitted supply of pH correctives. *K. marxianus* definitely metabolizes urea, and whey was repeatedly supplemented with urea as an N source for this yeast [28],[31],[32],[37],[45]-[47]. However, the effect of added urea on growth was often not described [37],[45],[46], or the published results were contradictory. Yadav et al. [28] observed an increased biomass yield, Kar and Misra [32] found no positive effect, Mahmoud and Kosikowski [31] described slight inhibition by added urea, and Rech et al. [47] even detected strong inhibition of growth which had been attributed to alkalinization. These inconsistent results require clarification by more detailed studies on this subject.

The first objective of this work was testing the effect of $(NH_4)_2SO_4$, $(NH_4)_2CO_3$, or urea as sources of nitrogen at aerobic growth of *K. marxianus* DSM 5422 in whey-borne medium with special attention on yeast growth and acidification. When using $(NH_4)_2SO_4$, sulfate remains in the medium and causes proton imbalance and acidification but, when using $(NH_4)_2CO_3$, carbonate disappears in form of $CO_2$ during cultivation which possibly avoids acidification. The second objective was studying the growth of *K. marxianus* DSM 5422 in whey-borne medium with urea in detail at defined conditions to get a deeper insight into the urea metabolism. The third objective was testing the pH-controlled feed of ammonia during aerobic growth of *K. marxianus* DSM 5422 in whey-borne medium. Ammonia as the cheapest source of nitrogen is interesting for large-scale processes.

# METHODS

## Microorganism

*K. marxianus* DSM 5422 from the Deutsche Sammlung von Mikroorganismen und Zellkulturen GmbH (Braunschweig, Germany) was maintained with the Cryoinstant preservation system (Lomb Scientific Pty Ltd, Vienna, Austria), cultivated on yeast-glucose-chloramphenicol agar (Roth GmbH, Karlsruhe, Germany) for 2 days at 32°C, and then used as an inoculum.

## Non-supplemented DW medium

All media originate from concentrated and partially demineralized sweet whey. Sweet whey was ultrafiltrated, concentrated by reverse osmosis, and then demineralized by slight alkalinization and moderate heating to yield the used whey permeate (thus processed in the Sachsenmilch Leppersdorf GmbH, Leppersdorf, Germany). Non-supplemented DW medium was prepared in batches of 1 L by mixing 0.5 L whey permeate with the same volume water. This mixture was autoclaved in a sealed 1-L Schott bottle for 15 min at 121°C. Precipitated minerals were allowed to settle overnight before the upper phase was withdrawn for cultivation experiments. Non-supplemented DW medium should not be confused with DW basic medium (as used in [11]-[14],[16]) which contains 10 g/L $(NH_4)_2SO_4$.

## Shake-Flask Cultivation

In a first series of shake-flask experiments, various media were prepared like the just-described non-supplemented DW medium with the modification that various sources of nitrogen (10 g/L $(NH_4)_2SO_4$, 10 g/L $(NH_4)_2CO_3$, or 5 g/L urea) were added before medium sterilization.

The media handling (autoclaving, settlement, withdrawal) occurred as in the above-described manner. These media were supplemented with 2 mL/L autoclaved trace-element solution (preparation as described in [10]) and diluted with water to 1/20 for reducing the content of sugar to 3.9 g/L and, thus, to avoid oxygen-limited growth at a low oxygen-transfer rate in shake flasks. Several 500-mL conical flasks with cotton plugs were filled each with 50 mL medium, inoculated with an agar-plate culture as described in [14], and shaken with 250 rpm at 32°C. After 14 h of pre-cultivation, the process was followed by regular analysis of the optical density (OD) and the pH value over a period of at least 10 h. After a total of 40 h of cultivation, the OD and pH were measured and the residual sugar and formed biomass were analyzed.

A second series of shake-flask experiments was based on autoclaved non-supplemented DW medium. Portions of this medium were supplemented with 2 mL/L sterile trace-element solution and separately autoclaved aqueous stock solutions of $(NH_4)_2SO_4$, $(NH_4)_2CO_3$, or urea. The pH was then adjusted to 6.5 with 0.1 M HCl or KOH. These media were diluted with water to reduce the sugar content to 3.9 g/L like in the first series. Discrete sterilization of whey and N sources in this second series avoided unwanted interaction between whey constituents and N compounds such as the Maillard reaction. Inoculation, cultivation, and analyses occurred as in the first series.

## Bioreactor cultivation

One reference experiment was done with DW basic medium (DW medium with 10 g/L $(NH_4)_2SO_4$) and described in detail in [11] while all other presented bioreactor processes were based on non-supplemented DW medium. The sterile 1-L stirred bioreactor with 0.6 mL Antifoam A (Fluka, Sigma Aldrich, St. Louis, USA) was filled with 0.6 L DW medium and 1.2 mL trace-element solution (the latter described in [10]). Further supplements such as $(NH_4)_2SO_4$, urea, and $Na_2SO_4$ were put into 100-mL Schott bottles, autoclaved for 15 min at 121°C, and then transferred to the bioreactor as follows: the bottle was connected with the bioreactor, some culture medium was pumped to the bottle, and, after complete dissolution of the substance, transported back to the reactor. Separate autoclaving avoided unwanted interaction such as the Maillard reaction and sulfate precipitation.

The cultivation occurred as usual [10]-[12],[14]: the medium was inoculated with five loops of biomass, and the reactor was operated at 1,200 rpm and 32°C and gassed with 50 L/h dry and $CO_2$-free air (given for 0°C and 101,325 Pa). The pH was controlled to ≥5 with 2 M KOH or 2 M ammonia solution. All processes lasted at least 36 h. Sampling and sample preparation were performed as usual [10].

## Test for Urea Assimilation

*K. marxianus* DSM 5422 was cultivated in diluted DW medium with urea (3.9 g/L sugars, 0.25 g/L urea) as described for the second series of shake-flask experiments. After depletion of the sugar, the cell suspension was separated into biomass and supernatant by centrifugation. The biomass was suspended in 50-mM phosphate buffer at pH 6 containing 0.25 g/L urea to yield an initial biomass concentration of 2 g/L, and the supernatant was supplemented with urea to give a concentration of 0.25 g/L. Both mixtures were then shaken with 250 rpm at 32°C. Sampling occurred at the beginning and after 24 h of incubation. The samples were analyzed regarding urea and ammonium.

## Analyses

The optical density of cell suspensions was measured photometrically at 600 nm (after pre-dilution to OD <0.4 if required). The biomass dry weight was determined by separating the yeasts via centrifugation, washing the pellet twice, and drying at 103°C. Sugar was quantified by a modified 3,5-dinitrosalicylic-acid method [48] with lactose as a standard. Ammonium, nitrate, nitrite, sulfate, and phosphate were measured by the LCK303, LCK339, LCK342, LCK153, and LCK049 cuvette tests (Hach Lange GmbH, Düsseldorf, Germany). The total dissolved nitrogen was analyzed by the Kjeldahl method (German DIN 38409 H11). Ethanol was quantified by gas chromatography [10].

Urea was measured in the style of Rahmatullah and Boyde [49] with some modifications. This method had been approved for quantifying urea in wine [50] being a matrix similar to culture liquids. Two reagents were prepared: a mixture of 300 mL sulfuric acid (95% to 98%, $\rho$ =1,840 g/L), 100 mL phosphoric acid (85%, $\rho$ =1,670 g/L), and 100 mL water; and a solution of 500 mg butane-2,3-dione monoxime (diacetyl monoxime) and 10 mg thiosemicarbazide in 100 mL water. The analysis reagent was prepared from 40 mL acid mixture and 20 mL of the second solution and used immediately. A volume of 0.1 mL sample was mixed with 3 mL analysis reagent. In the case of colored samples, this mixture was measured photometrically at 525 nm as a blind. Then, the mixture was put in a 25-mL test tube with plastic cap and incubated for 20 min in boiling water. After cooling and short mixing, the absorbance was measured at 525 nm. The blind was subtracted, and the obtained $\Delta A_{525nm}$ value was interpreted as a urea concentration by a non-linear calibration curve prepared for several solutions of urea in water (from 0 to 500 mg/L): $C_{Urea,L} = a \cdot (\Delta A_{525nm})^b$ with $a = 520$ mg/L and $b = 1.37$.

## Data analyses

The $O_2$ consumption, $CO_2$ formation, and the respiratory quotient (RQ) were calculated as usual[11]. The time-dependent specific growth rate was calculated from masses of formed $CO_2$ assuming a correlation between yeast growth and $CO_2$ formation which only applies to respiratory processes without significant maintenance: $\mu(t) \approx \Delta \ln(m_{CO2}(t))/\Delta t$. The overall biomass yield ($Y_{X/S}$) is the ratio between the mass of yeasts grown and the mass of sugar consumed; determination of these masses took losses by sampling and changes of the liquid volume into account. The pH-controlled feeding of 2 M KOH is given as a specific volume, related to the initial liquid volume of the culture. The mass of fed KOH was related to the formed biomass (given as $g_{KOH}/g_X$). The fed 2 M ammonia solution is also expressed as a specific volume and as the mass of supplied ammonia-N. The mass of bioavailable nitrogen was calculated as the sum of initial ammonium-N and urea-N plus the mass of ammonia-N supplied till a given time: $mN(t)=V_L(t_0) \cdot (CNH_4-N(t0)+C_{Urea}-N(t_0))+{}^mNH_4OH-N(t)$. The consumed nitrogen is this $m_N(t)$ value minus the nitrogen not yet used or lost by sampling:

$\Delta m_N(t)=m_N(t)-V_L(t) \cdot (CNH_{4-N}(t)+CUrea-N(t))-\Sigma m_N(sampling)$. These $\Delta m_N(t)$ values were used for calculating biomass-specific consumption rates ($r_N$ as $mg_N/g_X/h$). The overall biomass yield for nitrogen ($Y_{X/N}$) is the ratio between the mass of yeasts totally formed and the mass of N altogether consumed. The final N content of the biomass ($x_N$) is the reciprocal of this overall $Y_{X/N}$ value.

# RESULTS AND DISCUSSION

## Nitrogen in Non-Supplemented DW Medium

The composition of whey depends on many factors such as origin of milk (cow, goat, or sheep), technology of curd production, and whey processing (e.g., [51],[52]). Casein protein is coagulated by acidification (mineral or organic acids directly added or lactic acid produced in the processed milk by bacteria) and/or by using chymosin. Whey processing modifies the composition of whey as well: whey protein is separated by ultrafiltration, solutes are concentrated by reverse osmosis, and/or minerals are partially removed by alkalinization. Preparation of whey-borne culture media also changes the composition by dilution, adding supplements, and heat sterilization. This explains why published compositions of whey-borne media highly fluctuate.

Several batches of non-supplemented DW medium were analyzed regarding potential sources of nitrogen and some other parameters (Table 1). The medium is rich in \sugar and thus exhibits a high potential for biomass growth. An amount of 78 g/L sugar allows formation of 28 g/L *K. marxianus* biomass assuming $Y_{X/S} = 0.36$ g/g (as observed at aerobic cultivation in DW basic medium with trace elements [11]). These 28 g/L biomass can only develop when yeast growth is not limited by any other resource than sugar. *K. marxianus* DSM 5422 grown under such conditions exhibited the following content of minerals (in milligrams of the addressed element per gram dry biomass): $x_N = 78...79$ mg/g [10],[11], $x_P = 10$ mg/g, $x_S = 4$ mg/g, and $x_K = 2$ mg/g (unpublished results). These data are similar to published compositions of *K. marxianus*[45],[53]-[55] and*Saccharomyces cerevisiae*[40]. Multiplying these *x* values with the cell concentration of 28 g/L gives the required concentration of the respective element in the culture medium to allow non-limited growth: 2,200 mg/L N, 280 mg/L P, 112 mg/L S, and 56 mg/L K. Nitrogen and sulfur are lacking in non-supplemented DW medium while the phosphorous and potassium content covers the demand (Table 1).

**Table 1:** Composition of non-supplemented DW medium

| Composition of non-supplemented DW medium | |
|---|---|
| Parameter | Value |
| Sugars (88% lactose, 12% galactose) | 78 g/L |
| Total N content (Kjeldahl analysis) | 403 mg/L |
| Ammonium-N | 106 mg/L |
| Urea-N | 43 mg/L |
| Nitrate-N | 4 mg/L |
| Nitrite-N | <1 mg/L |
| Organic nitrogen (except urea-N) | 250 mg/L |
| Proteina | 1,560 mg/L |
| Sulfate-S | 78 mg/L |
| Phosphate-P | 410 mg/L |
| Potassium | 2,600 mg/L |
| pH value | 5.7 to 6.0 |

Concentrated and partially demineralized sweet whey permeate was diluted with the same volume of water and autoclaved for 15 min at 121°C; precipitates were settled, and the upper clear phase was analyzed.

[a]According to the EC Nutrition Labelling Rules Directive 90/496/EEC, organic N multiplied by 6.25.

Löser et al.

Löser et al. Energy, Sustainability and Society 2015 **5**:2  doi:10.1186/s13705-014-0028-2

DW medium owns 403 mg/L Kjeldahl-N (Table 1). Detailed analysis of five batches of non-supplemented DW medium gave 89 to 130 mg/L ammonium-N (106 mg/L on an average, $\sigma_{n-1} = 15$ mg/L), 38.4 to 45.9 mg/L urea-N (43 mg/L on an average, $\sigma_{n-1} = 2.6$ mg/L), 4 mg/L nitrate-N, and <1 mg/L nitrite-N. The Kjeldahl-N minus the urea-N and inorganic N gives a proteinogenic N of 250 mg/L which corresponds to ca. 1.5 g/L proteins (Table 1).

Utilization of whey proteins by *K. marxianus* is discussed controversially. Raw cheese whey owns 7 g/L protein comprising

50% β-lactoglobulin, 20% α-lactalbumin, 15% glycomacropeptide, and 15% minor protein/peptide components [51]. Their microbial hydrolysis requires excretion of proteases. Decomposition of whey proteins by K. marxianus has been repeatedly studied; the results varied from absent hydrolysis [26], over 20% to 33% [28],[29],[56], up to 80% hydrolysis[33]. Recent studies confirmed an extracellular serine protease for K. marxianus[57], and Yadav et al. [28] proved modification of whey proteins by K. marxianus via electrophoresis. Indigenous proteases in milk [58] and the proteolytic activity of lactobacilli [56] could also contribute some to protein modification during milk processing. Pre-treatment of whey protein with added proteases resulted in peptides <1 kDa which were efficiently assimilated by K. marxianus[59].

Some yeasts and fungi assimilate nitrate by intracellular reduction to ammonium [60]. K. marxianus seems to be unable for nitrate assimilation; at least several tested strains were negative [61]. This explains why nitrate added to whey did not improve growth of K. marxianus[32]. Some growth of K. marxianus in $NaNO_3$-supplemented medium was possibly caused by the added yeast extract [62]. K. marxianus DSM 5422 proved to be unable to assimilate nitrate.

Due to the uncertainty of whey-protein assimilation by K. marxianus, it is assumed that whey proteins do not contribute to assimilable nitrogen. The utilizable nitrogen in non-supplemented DW medium of ca. 150 mg/L (ammonium-N plus urea-N) is thus much smaller than the calculated demand. Supplementation of DW medium with nitrogen is essentially required.

K. marxianus DSM 5422 grows well with ammonium [10],[11], but its growth with urea has not yet been tested. Yeasts generally assimilate urea [44] which should also apply to the studied strain.

A low amount of iron, zinc, and copper in DW medium limits growth of K. marxianus DSM 5422[11]. Here, the medium was ever supplemented with trace-element solution to avoid such limitation.

# Test of Several Sources of Nitrogen without Previous pH Adjustment

This preliminary test for assimilation of diverse sources of nitrogen by *K. marxianus* DSM 5422 was performed in shake flasks. Non-supplemented DW medium was spiked with several N sources (Table 2): no supplement as a reference, $(NH_4)_2SO_4$ as the usual N supplement, $(NH_4)_2CO_3$ as an alternative ammonium resource, and urea as another N compound. The DW medium contributed ca. 7 mg/L N (5 mg/L $NH_4$-N and 2 mg/L urea-N) to the assimilable N in each culture. The initial pH value was influenced by the added N source: $(NH_4)_2SO_4$ did not change the pH while urea and $(NH_4)_2CO_3$ alkalinized the medium (Table 2). All shake flasks were inoculated and pre-cultivated for 14 h before the process was followed by repeated OD and pH measurements (Figure 1).

**Table 2:** Parameters at aerobic growth of *K. Marxianus* DSM 5422 in media with various sources of nitrogen without previous pH adjustment

| Parameter | | Added source of nitrogen | | | |
|---|---|---|---|---|---|
| | | No supplement | (NH4)2SO4 | (NH4)2CO3 | Urea |
| Supplied N | [mg/L] | 0 | 106 | 146 | 117 |
| NH4-N + urea-N | [mg/L] | 7 | 113 | 153 | 124 |
| Initial pH value | [−] | 5.96 | 5.94 | 7.50 | 7.18 |
| μ at t = 14 h | [h−1] | 0.51 | 0.57 | 0.29 | 0.57 |
| Final sugar | [g/L] | 0.92 | 0.00 | 0.15 | 0.00 |
| Final biomass | [g/L] | 0.47 | 1.23 | 1.06 | 1.29 |
| Final pH value | [−] | 3.99 | 2.47 | 3.45 | 7.10 |

The whey and the sources of nitrogen were autoclaved together; the data belong to the experiment shown in Figure 1; the final parameters were measured after a total of 40 h of cultivation.

Löser *et al.*

Löser *et al. Energy, Sustainability and Society* 2015 **5**:2, doi:10.1186/s13705-014-0028-2

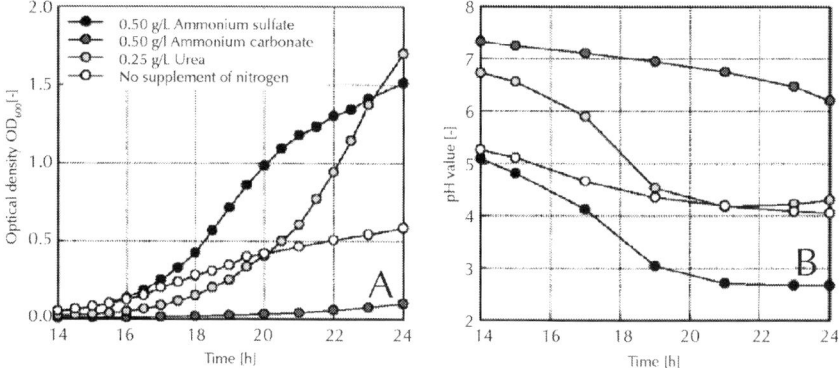

**Figure 1:** Aerobic batch cultivation of*K. marxianus*DSM 5422 without pH adjustment. Aerobic batch cultivation of *K. marxianus* DSM 5422 in media with various sources of nitrogen without previous pH adjustment. Concentrated and partially demineralized sweet whey was diluted with the same volume of water, supplemented with various sources of nitrogen, and autoclaved; the clear upper phase was supplemented with trace elements and diluted again with water resulting in media with 3.9 g/L sugar; the media were inoculated and cultivated in conical flasks at 250 rpm and 32°C.

The process without a supplement was at first similar to the process with $(NH_4)_2SO_4$ (Figure 1), but later, nitrogen became a limiting factor so that growth and medium acidification slowed down and nearly stopped after 24 h. The N-limited growth resulted in some residual sugar (Table 2).

With a supplement of $(NH_4)_2SO_4$, *K. marxianus* DSM 5422 grew at first with $\mu = 0.57$ h$^{-1}$ as usual in diluted DW medium [12],[14], but then, the growth slowed down due to medium acidification (Figure 1). A pH <3.5 impairs growth of *K. marxianus*[28],[42],[43]. This acidification was caused by consumption of ammonium without an equivalent uptake of sulfate. The use of ammonium sulfate as an N source calls for buffered medium (in shake-flask experiments [12],[14]) or for pH control (in bioreactor experiments [10],[11],[13],[16],[22]).

The supplement of $(NH_4)_2CO_3$ at first alkalinized the medium. The initial pH of 7.5 impacted growth of *K. marxianus* DSM 5422 and caused a highly retarded process (Figure 1). Vivier et al. [42] and Antoce et al. [43] found that a pH >7 is adverse for growth of *K. marxianus*. Continued growth gradually reduced the pH and accelerated yeast growth so that all sugar had been consumed after 40 h. Carbonate as an exchangeable anion can leave the medium in the form of carbon dioxide which counteracted acidification (medium with $(NH_4)_2CO_3$) while sulfate as a permanent ion remains and resulted in unfavorable acidification (medium with $(NH_4)_2SO_4$).

The supplement of urea also alkalinized the medium to some degree (initial pH = 7.18) which is, at the first view, surprising since urea reacts neutral in aqueous solution. Autoclaving the medium together with urea maybe caused some hydrolysis of urea to form alkaline ammonium carbonate. This slight alkalinization slowed down yeast growth and retarded the process a little at the beginning, but the yeast metabolism lowered the pH so that the growth rate approached a normal value during pre-cultivation (Table 2). After 21 h, the medium acidification stopped and then the pH increased to a final pH of 7.1 (i.e., only temporary decrease in pH). Absence of enduring pH changes confirms the earlier postulated proton neutrality at growth with urea [38]-[40].

# Test of Several Sources of Nitrogen with Previous pH Adjustment

The just-described shake-flask experiments were repeated with two modifications: all sources of nitrogen were autoclaved separately to eliminate unwanted interaction between the N sources and medium constituents, and the initial pH value was adjusted to 6.5 for avoiding alkaline conditions.

The convenient initial pH of 6.5 produced a uniformly high specific growth rate of $\mu = 0.59$ h$^{-1}$(Table 3) resulting in very similar cell densities at the beginning of the observation period (Figure 2A). This $\mu$ value was identical with the growth rate of *K. marxianus* DSM 5422 in highly diluted and phosphate-buffered DW basic medium [12],[14].

**Table 3:** Parameters at aerobic growth of *K. marxianus*DSM 5422 in media with various sources of nitrogen with initial pH adjustment

| Parameter | | Added source of nitrogen | | | |
|---|---|---|---|---|---|
| | | No supplement | (NH4)2SO4 | (NH4)2CO3 | Urea |
| Supplied N | [mg/L] | 0 | 106 | 146 | 117 |
| NH4-N + urea-N | [mg/L] | 7 | 113 | 153 | 124 |
| Initial pH value | [−] | 6.50 | 6.50 | 6.50 | 6.50 |
| µ at t = 14 h | [h−1] | 0.59 | 0.59 | 0.59 | 0.59 |
| Final sugar | [g/L] | 1.03 | 0.00 | 0.00 | 0.00 |
| Final biomass | [g/L] | 0.47 | 1.29 | 1.25 | 1.44 |
| Final pH value | [−] | 4.29 | 2.43 | 2.54 | 6.45 |

The whey and the sources of nitrogen were autoclaved separately; the data belong to the experiment shown in Figure 2; the final parameters were measured after a total of 40 h of cultivation.

Löser *et al.*

Löser *et al. Energy, Sustainability and Society* 2015 **5**:2   doi:10.1186/s13705-014-0028-2

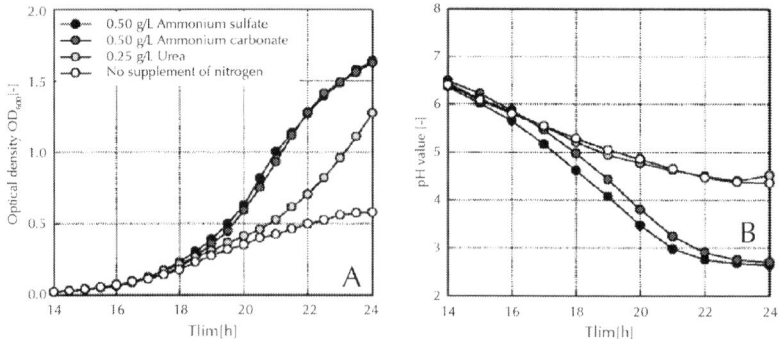

**Figure 2:** Aerobic batch cultivation of *K. marxianus*DSM 5422 with initial pH adjustment. Aerobic batch cultivation of *K. marxianus* DSM 5422 in media

with various sources of nitrogen with initial pH adjustment. Non-supplemented DW medium was supplemented with various sources of nitrogen and trace elements, the pH was adjusted to 6.5 by HCl or KOH, and these media were diluted with water resulting in 3.9 g/L sugar; the media were inoculated and cultivated in conical flasks at 250 rpm and 32°C.

The culture without a supplement of nitrogen exhibited N-limited growth (Figure 2A), only moderate acidification owing to restricted growth (Figure 2B), a low biomass formation, and some residual sugar after 40 h of cultivation (Table 3). Supplementation with $(NH_4)_2SO_4$ or $(NH_4)_2CO_3$ resulted in a nearly identical growth behavior and similar final pH values and biomass yields (Table 2); the pH was suboptimal in both cases which caused impaired growth compared to the process in phosphate-buffered medium [12],[14]. A reduced biomass yield for *K. marxianus* in whey at a low pH has also been described by Yadav et al. [28] and interpreted as diversion of lactose from anabolism (growth) toward catabolism (maintenance). The use of $(NH_4)_2CO_3$ instead of $(NH_4)_2SO_4$ was without advantage since the initial pH adjustment of $(NH_4)_2CO_3$-supplemented medium with hydrochloric acid caused substitution of exchangeable by permanent ions (carbonate replaced by chloride). $(NH_4)_2CO_3$ is thus not a useful alternative since a high initial pH inhibits yeast growth while preceding pH adjustment with acids eliminates the buffering effect of carbonate.

Separate autoclaving of the N source and initial pH adjustment caused good growth in the urea-supplemented medium from the beginning; then, the growth became somewhat retarded (at 18 to 22 h; Figure 2A), but afterward, the growth accelerated again (at $t > 22$ h) and even exceeded growth with $(NH_4)_2SO_4$ (Table 3). Quick growth in the first period was possibly based on whey-borne ammonium while the temporal slowdown of growth in the second period perhaps came about through the adaptation of yeast metabolism to urea assimilation after $NH_4$-N depletion. The quite high biomass yield with urea is certainly attributed to an appropriate pH value over the whole growth period (compare [39],[62]). Urea is thus a promising source of nitrogen for *K. marxianus*. The acidification with urea was only of temporal nature; the pH increased again to give a final value being nearly identical with the initial pH (due to proton neutrality at assimilation of urea [38]-[40]).

# Assimilation of Urea by *K. Marxianus* DSM 5422

Ammonium is utilized by all common yeasts directly while urea is either hydrolyzed by urease to form ammonium or it is assimilated via the urea amydolyase pathway [25],[38]. Urease acts extracellularly while the amydolyase pathway works intracellularly. *K. marxianus* is regarded as a urease-negative yeast [40] and should metabolize urea only in the latter way being a two-step process [44],[60],[63]; urea reacts with hydrocarbonate in an energy-consuming process to form allophanate which, in turn, is hydrolyzed to release ammonium:

$$NH2?CO?NH_2 + HCO_3^- + ATP \rightarrow NH_2?CO?NH?COO^- + H_2O + ADP + Pi \quad (1)$$

$$NH_2 ?CO?NH?COO^- + 3\ H2O + H^+ \rightarrow 2\ NH_4^+ + 2\ HCO_3^- \quad (2)$$

These two reactions are catalyzed by urea carboxylase and allophanate hydrolase [44],[60],[63]. The produced ammonium is then metabolized in the same manner as ammonium taken up directly.

An experiment was done for clarifying the way of urea assimilation in *K. marxianus* DSM 5422. The yeast was cultivated in diluted DW medium with urea, and the obtained culture was then used in an above-described test. *K. marxianus* DSM 5422 did not excrete urease into the medium since urea added to the cell-free aqueous fraction of this culture was not hydrolyzed at all. This is in accordance with Nahvi and Moeini [61] who found all tested *K. marxianus* and *K. lactis* strains being urease negative. These findings argue for assimilation of urea via the amydolyase pathway. The grown yeast biomass was incubated with urea in a phosphate buffer but only a bit urea was reacted to ammonium (average reaction rate 0.25 mg urea/$g_x$/h). Absent sugar obviously suppressed transfer of urea into ammonium which refers to an amydolyase pathway being under transcriptive control.

# Bioreactor Cultivation with the Addition of Urea

The shake-flask experiments clearly demonstrated the capability of *K. marxianus* DSM 5422 for urea assimilation. Urea is a promising source of nitrogen due to the proton neutrality at its consumption during yeast growth [38]-[40], avoiding strong acidification as happening at growth with $(NH_4)_2SO_4$. Urea has been repeatedly used at growth of *K. marxianus*, but the obtained results were inconsistent [28],[31],[41],[47],[62]. Hensing et al. [39] referred to the potential risk of an imbalance between release and assimilation of ammonium; the medium could alkalinize when ammonium is quicker released from urea than is incorporated into biomass. Such an alkalinization was observed during cultivation of *K. marxianus* in urea-supplemented whey [47]. In the above-described experiments, the pH temporally decreased rather than increased. This phenomenon has not yet been understood and requires clarification in a bioreactor experiment.

*K. marxianus* DSM 5422 was cultivated in a stirred reactor at well-defined conditions (32°C, $pO_2 \geq 30\%$ air saturation, pH $\geq 5$) in DW medium which was supplemented with urea, $Na_2SO_4$, and trace-element solution to avoid limitation of growth by nitrogen, sulfur or microelements. Another bioreactor experiment performed with DW medium containing 10 g/L $(NH_4)_2SO_4$ and trace elements was taken from [11] and used here as a reference. These two processes are depicted in Figure 3, and characteristic parameters are summarized in Table 4.

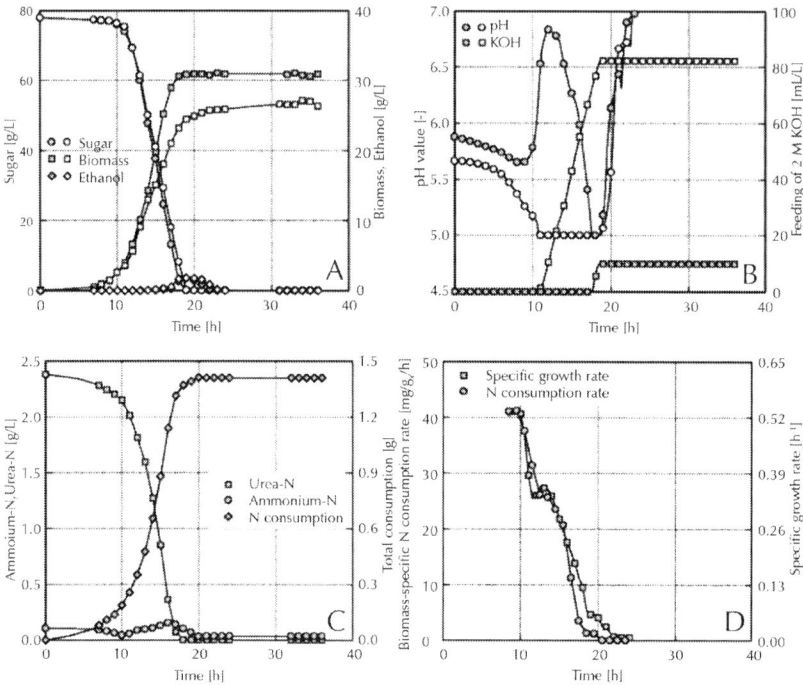

**Figure 3:** Aerobic batch cultivation of *K. marxianus*DSM 5422 in stirred bio-reactors using ammonium or urea. Aerobic batch cultivation of *K. marxianus* DSM 5422 in a stirred bioreactor using ammonium (white symbols) or urea (grey symbols) as a source of nitrogen. DW medium with 2 mL/L trace-ele-ment solution was supplemented with $(NH_4)_2SO_4$ or urea plus $Na_2SO_4$ (white symbols = 10 g/L $(NH_4)_2SO_4$; grey symbols = 5 g/L urea and 0.4 g/L $Na_2SO_4$); the cultivation occurred in an 1-L stirred reactor at 1,200 rpm, 32°C, and aeration with 50 L/h; the pH was controlled to ≥5 with 2 M KOH; the given growth rate was derived from measured $CO_2$ data.

**Table 4:** Parameters at aerobic growth of *K. marxianus*DSM 5422 in media with various sources of nitrogen in a stirred bioreactor

| Parameter | | Added sources of nitroge | | | | |
|---|---|---|---|---|---|---|
| | | (NH4)2SO4a | Urea | NH4OH | (NH4)2SO4+ NH4OH | (NH4)2SO4+ NH4OH |
| Added (NH4)2SO4 | [g/L] | 10 | 0 | 0 | 1 | 2 |
| Added urea | [g/L] | 0 | 5 | 0 | 0 | 0 |

| Added Na2SO4 | [g/L] | 0 | 0.4 | 0.4 | 0 | 0 |
|---|---|---|---|---|---|---|
| Initial NH4-N content | [g/L] | 2.12 | 0.10 | 0.13 | 0.30 | 0.53 |
| Initial urea-N content | [g/L] | 0.04 | 2.38 | 0.04 | 0.04 | 0.04 |
| Maximum growth rate | [h−1] | 0.56 | 0.54 | 0.54 | 0.54 | 0.56 |
| Consumed sugar | [g] | 45.2 | 44.7 | 44.3 | 44.8 | 44.9 |
| Final cell concentration | [g/L] | 26.7 | 30.8 | 7.49 | 22.1 | 27.0 |
| Formed biomass | [g] | 16.3 | 17.3 | 4.2 | 13.0 | 16.2 |
| Overall YX/S | [g/g] | 0.362 | 0.388 | 0.095 | 0.289 | 0.361 |
| Final NH4-N content | [g/L] | 0.02 | 0.03 | 0.00 | 0.04 | 0.17 |
| Final urea-N content | [g/L] | 0.00 | 0.00 | 0.00 | 0.00 | 0.00 |
| Supplied 2 M NH4OH | [mL] | 0.0 | 0.0 | 0.0 | 24.8 | 39.8 |
| Supplied ammonia-N | [g] | 0.000 | 0.000 | 0.000 | 0.694 | 1.114 |
| Consumed NH4-N | [g] | 1.212 | 0.039 | 0.077 | 0.849 | 1.317 |
| Consumed urea-N | [g] | 0.024 | 1.371 | 0.025 | 0.026 | 0.023 |
| Consumed N | [g] | 1.236 | 1.410 | 0.102 | 0.875 | 1.340 |
| Overall YX/N | [g/g] | 13.2 | 12.3 | 41.2 | 14.9 | 12.1 |
| Final N content, xN | [mg/g] | 76 | 81 | 24 | 67 | 83 |
| Supplied 2 M KOH | [mL/L] | 82 | 10 | 0 | 0 | 0 |
| Supplied KOH | [gKOH/ gX] | 0.34 | 0.04 | 0.00 | 0.00 | 0.00 |
| Consumed oxygen | [g] | 25.3 | 26.4 | 28.6 | 30.3 | 27.9 |
| Formed CO2 | [g] | 35.1 | 37.2 | 41.2 | 42.5 | 39.0 |
| Average RQ | [mol/ mol] | 1.01 | 1.02 | 1.05 | 1.02 | 1.02 |
| Maximum CEtOH,L | [g/L] | 1.8 | 1.8 | <0.1 | 0.2 | <0.1 |

Final parameters are valid for 36 h of cultivation with exception of the process with $NH_4OH$ as the only added N source where the final parameters were determined for 40 h of cultivation; cultivation at conditions as stated in Figures 3 and 4.

[a] Data in this column was in part taken from Urit et al. [10],[11],[14]; small deviations from the formerly published data are due to re-evaluation of basic data taking liquid sampling into account.

Löser *et al.*

Löser *et al. Energy, Sustainability and Society* 2015 **5**:2   doi:10.1186/s13705-014-0028-2

The courses of yeast growth, sugar consumption, and marginal ethanol formation were very similar for $(NH_4)_2SO_4$ or urea as the added N sources (Figure 3A). The only marked difference was the amount of formed biomass which was apparently higher with urea (Table 4). This observation is in accordance with Hensing et al. [39] and Rajoka et al. [62]. From the energetic point of view, growth with urea should be less effective compared to growth with ammonium since assimilation of urea via the amydolyase pathway requires ATP [44],[60],[63]. Cultivation with urea should therefore result in a lower rather than a higher biomass yield. In case of Hensing et al. [39] and Rajoka et al. [62], the observed low growth with ammonium was possibly caused by inhibitory acidification due to absent pH control at shake-flask cultivation. Such an inhibitory acidification was prevented by controlling the pH during bioreactor cultivation (Figure 3B); here, the diverging biomass yields were presumably caused by slightly different amounts of bioavailable nitrogen (1.49 g N in medium with 5 g/L urea, and 1.30 g N in medium with 10 g/L $(NH_4)_2SO_4$). Nearly all bioavailable nitrogen was assimilated in both processes (only 20 or 30 mg/L residual $NH_4$-N in the culture broth; Table 4) which refers to a slight deficit of nitrogen, and this deficit was more striking with $(NH_4)_2SO_4$. This argumentation is supported by a higher N content in the biomass grown with urea (Table 4).

In urea-supplemented medium, *K. marxianus* DSM 5422 grew at first quickly; but later, the growth slowed down more and more (Figure 3D), although sufficient O, N, P, S, K, and trace elements should allow non-limited growth over an extended period. This behavior could be explained with a lack of vitamins in whey [27],[29] or with inhibition by whey-borne minerals [31]. Supplementing whey with yeast extract [27],[29] or vitamins [29],[32],[47] stimulated the growth of *K. marxianus*.

The source of nitrogen distinctly influenced the pH($t$) course (Figure 3B). With $(NH_4)_2SO_4$, the pH decreased due to ammonium consumption until the pH was controlled to pH 5; 82 mL/L 2 M KOH were supplied which corresponds to a specific dosage of 0.34 g KOH per g of produced biomass. The strong acidification with $(NH_4)_2SO_4$ is explained by a proton imbalance [38]-[40]: ammonium was consumed while most of the sulfate remained in the medium. With urea, the pH at first a little decreased, then sharply rose to pH 6.8 and reduced again to pH 5 where the pH controller avoided further acidification; the KOH dosage was in fact much smaller (only 10 mL/L 2 M KOH or 0.04 g KOH per g of produced biomass; Table 4). Hensing et al. [39] cultivated K. lactis in galactose medium with $(NH_4)_2SO_4$ or urea; the acidification was strong and permanent with $(NH_4)_2SO_4$, while the acidification was moderate and only temporary with urea.

During cultivation of K. marxianus DSM 5422 in urea-supplemented DW medium, the dissolved ammonium-N and urea-N were repeatedly measured and used for calculating the N consumption (Figure 3C). Intracellular conversion of urea to ammonium and usage of this ammonium for biomass growth occurred with nearly the same rate since ammonium excretion was only marginal (some $NH_4$-N originated from the used whey). The N consumption correlated well with the yeast growth (compare Figure 3A and C) and, thus, the courses of the N consumption rate and the specific growth rate were similar (Figure 3D). The quotient of these rates represents a momentary $Y_{X/N}$ value.

The small transient $NH_4$-N accumulation (Figure 3C) partially correlated with the temporary increase in pH. Hensing et al. [39] already referred to the danger of alkalinization when ammonium release exceeds ammonium assimilation. Such an alkalization to pH 8.5 was observed by Rech et al. [47] at cultivation of K. marxianus in urea-supplemented whey causing severe growth inhibition. Here, such an inhibition did not occur (only moderate rise of pH to 6.8).

# Bioreactor Cultivation at a pH-Controlled Feed of Ammonia

Feeding the required nitrogen in form of ammonia could be a cost-saving alternative. Ammonia was repeatedly used as an N source at cultivation of K. marxianus in whey or other media [34],[64]-[67] but dissolved ammonium or N consumption has not been paid much attention, with exception of Hack and Marchant [65] who depicted the time-dependent supply of ammonia.

In another series of bioreactor experiments, *K. marxianus* DSM 5422 was cultivated in DW medium as before but 2 M $NH_4OH$ was used as the predominating N source which was supplied by the pH controller at pH <5. Three experiments were performed with a varied mass of $(NH_4)_2SO_4$ which was added as a pure substance to the autoclaved DW medium (0, 0.6, or 1.2 g). These processes were limited neither by oxygen ($pO_2$ always >10%) nor by sulfur (proven by residual sulfate).

In DW medium without an $(NH_4)_2SO_4$ supplement (Figure 4; white symbols), *K. marxianus* DSM 5422 grew on whey-borne nitrogen (0.13 g/L $NH_4$-N and 0.04 g/L urea-N), but this nitrogen was quickly depleted (Figure 4C) and the growth became N limited (Figure 4A). The pH temporally rose (Figure 4B) which seemingly correlated with urea consumption (Figure 4C). After depletion of all bioavailable N, the pH decreased only slowly due to a low metabolic activity (look at the sugar concentration in Figure 4A). Later, the pH stagnated above pH 5 and ammonia was thus not dosed (Figure 4B). The low availability of nitrogen (Figure 4D) caused restricted yeast growth (Figure 4A). The total N consumption and the N consumption rate were accordingly low (Figure 4E,F).

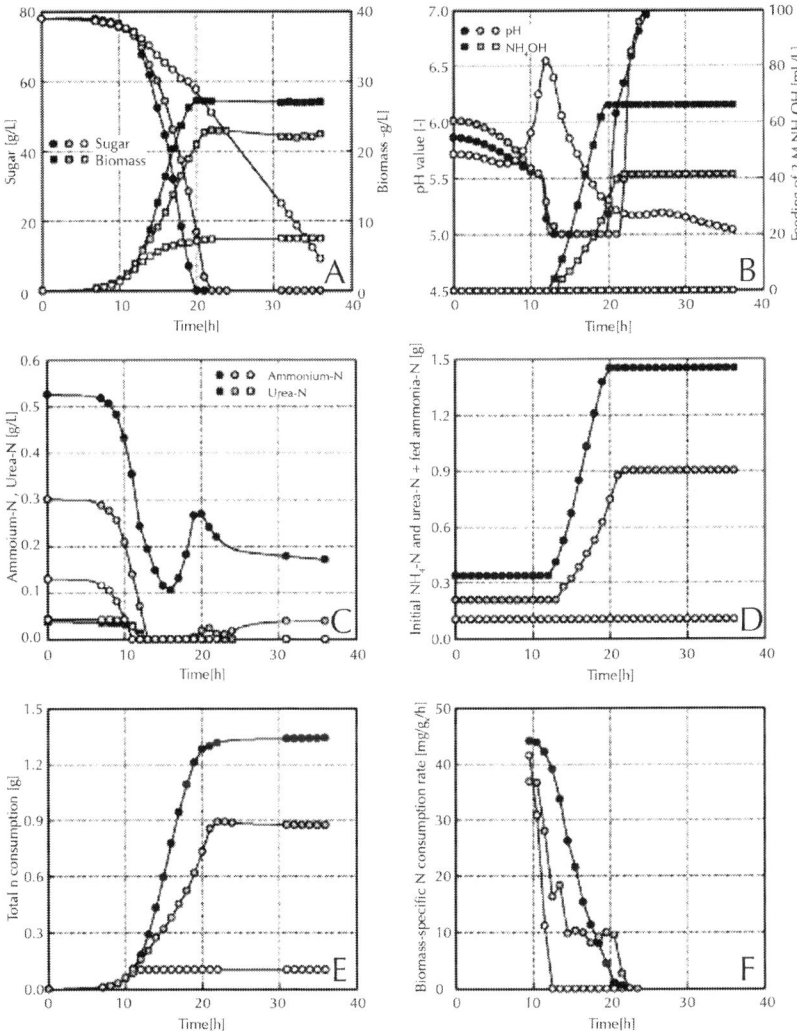

**Figure 4:** Aerobic batch cultivation of *K. marxianus*DSM 5422 in stirred biore-actors using (NH4)2SO4and NH4OH. Aerobic batch cultivation of *K. marxianus* DSM 5422 in a stirred bioreactor using various amounts of $(NH_4)_2SO_4$ and $NH_4OH$ as sources of nitrogen. DW medium with 2 mL/L trace-element solution was supplemented with $(NH_4)_2SO_4$ or $Na_2SO_4$ (white symbols = 0.4 g/L $Na_2SO_4$; grey symbols = 1 g/L $(NH_4)_2SO_4$; black symbols = 2 g/L $(NH_4)_2SO_4$); the cultivation occurred in an 1-L stirred reactor at 1,200 rpm, 32°C, and aeration with 50 L/h; the pH was controlled to ≥5 with 2 M $NH_4OH$; the given growth rates were derived from measured $CO_2$ data.

A supplement of 1 g/L $(NH_4)_2SO_4$ in the second experiment (Figure 4; grey symbols) increased the bioavailable N (Figure 4D) and allowed better yeast growth due to the higher initial $NH_4$-N which let the pH quickly decrease (Figure 4B). After depletion of this nitrogen, the pH stagnated at 5.05 for a while and then dropped below 5 where the pH controller started dosage of ammonia (Figure 4B). The added ammonia was assimilated immediately, and no ammonium accumulated in the medium (Figure 4C). That is, the yeast growth continued but at N-limited conditions as becoming visible from the low rates of ammonia dosage and N consumption (Figure 4D,F). This deficit of nitrogen slowed down growth and diminished the formed biomass (Figure 4A, Table 4). In the stationary period, some ammonium was released into the medium (Figure 4C) which was also observed by Ghaly and Kamal [26] at the cultivation of *K. marxianus* in whey and interpreted as decomposition of yeast biomass with release of $NH_4$-N into the medium.

A supplement of 2 g/L $(NH_4)_2SO_4$ in the third experiment (Figure 4; black symbols) increased the bioavailable N most (Figure 4D) and resulted in fast yeast growth and quick acidification. The feed of ammonia at pH <5 started before the initially added $NH_4$-N had been depleted (Figure 4C). The early start of ammonia dosage prevented limitation of yeast growth by nitrogen ($NH_4$-N always >100 mg/L), and the supply and uptake of nitrogen were well balanced (Figure 4C). The high rate of ammonia dosage corresponded with an accordingly fast growth and intensive N consumption (Figure 4D,E). The rate of N consumption became gradually smaller which is explained by the gently declining growth rate (possible reasons for this fading growth were discussed above). Urea was assimilated co-metabolically with the ammonium (Figure 4C).

The process with a supplement of 2 g/L $(NH_4)_2SO_4$ and ammonia dosage ran very similar to the process with 10 g/L $(NH_4)_2SO_4$; the final cell concentrations, formed biomasses, and the overall $Y_{X/S}$ values were nearly identical in both processes (Table 4) which demonstrates effective cultivation of *K. marxianus* with a pH-controlled feed of ammonia. Supplementing the medium with some $(NH_4)_2SO_4$ was however required for a quick acidification and for initiation of ammonia dosage. The amount of added $(NH_4)_2SO_4$ could be reduced by changing the setpoint of the pH controller (e.g., to pH 5.5) so that dosage of ammonia starts earlier and avoids N-limited conditions even at a reduced $(NH_4)_2SO_4$ supplement, but some $(NH_4)_2SO_4$ is needed to cover the requirement for sulfur.

# Nitrogen in Biomass

The content of nitrogen in biomass grown at cultivation in the stirred bioreactor (Figures 3 and 4) is the inverse of the overall biomass yield for nitrogen: $x_N = 1/Y_{X/N}$. The overall $Y_{X/N}$ values were calculated from the produced biomass and the consumed nitrogen, assuming that only ammonium and urea were assimilated (Table 4). This calculation ignores that K. marxianus possibly hydrolyzes some whey-borne proteins and assimilates thus-formed peptides and amino acids. The N content of biomass depended on the extent of N limitation (Table 4): cultivation with enough nitrogen (process with 5 g/L urea and process with 2 g/L $(NH_4)_2SO_4$ plus dosed ammonia) gave $x_N$ values of 81 and 83 mg/g, a slight deficit of nitrogen during the late growth stage (process with 10 g/L $(NH_4)_2SO_4$) resulted in $x_N$ = 76 mg/g, distinct N limitation (process with 1 g/L $(NH_4)_2SO_4$ plus ammonia) yielded $x_N$ = 67 mg/g, while severe N limitation (process without any N supplement) produced an $x_N$ value of only 24 mg/g. A diminished N content of K. marxianus was also observed at limitation of growth by trace elements [10],[11]. A decreased N content can be explained by a lowered portion of active biomass owing to intracellular storage of polysaccharides (details in [10],[11]).

The N content can also be derived from the elemental composition of biomass. Several authors measured the cell composition for K. marxianus by elemental analyzers and transformed these data into biomass formulae: $CH_{1.78}O_{0.75}N_{0.16}$[45], $CH_{1.776}O_{0.575}N_{0.159}$[53], $CH_{1.63}O_{0.54}N_{0.16}$[54], $CH_{1.94}O_{0.76}N_{0.17}$[55]. These formulae represent an N content of 88, 80, 91, or 83 mg/g. The fluctuations originate from measuring errors and from a variable cell composition depending on growth conditions [68].

# Stoichiometry of yeast growth

Stoichiometric equations for describing the growth of K. marxianus has been derived here by using the method of Hensing et al. [39] and Mazutti et al. [69]. Such balancing requires a sum formula for biomass. The above-given formulae for *K. marxianus* biomass are restricted to C, H, O, and N as the predominating elements (derived from elemental analyses [45], [53]-[55]). Here, the elements P, S, and K are included for more precision. The C, H, and O content was taken from the above-given biomass formulae (as averages), the N content of 82 mg/g was taken from own measurements at non-limited yeast growth, and the P, S, and K content was assumed with 10, 4, and 2 mg/g (own assimilation measurements). Combination of these data gives $CH_{1.78}O_{0.66}N_{0.158}P_{0.009}S_{0.0035}K_{0.0015}$ (yielding a molar mass of 27.027 g/mol).

Individual stoichiometric balance equations were derived for ammonium, urea, or ammonia as an N source, assuming respiratory growth (without formation of ethanol or ethyl acetate) of *K. marxianus* DSM 5422 with lactose as a substrate. The included stoichiometry coefficients were determined by balancing each element: seven balance equations were obtained containing nine unknown stoichiometric coefficients. This uncertain algebraic system was dissolved following Hensing et al. [39] by adding a proton balance and introducing the yield coefficient ($Y_{X/S}$ informs about the relation of assimilatory to dissimilatory substrate utilization and allows to establish the mass ratio between formed biomass and consumed lactose). $Y_{X/S} = 0.36$ g/g was used here uniformly for all balances as found at non-limited growth with ammonium or ammonia (Table 4). Phosphate and sulfate were consumed in form of $HPO_4^{2-}$ and $SO_4^{2-}$ at the prevailing pH. Three equations were obtained for ammonium, ammonium hydroxide, or urea as an N source:

$$C_{12}H_{22}O_{11} + 0.0410\, HPO_4^{2-} + 0.0160\, SO_4^{2-} + 0.0068\, K^+ + 7.3790\, O_2 + 0.7205\, NH_4^+ \rightarrow 4.5600\, CH_{1.78}O_{0.66}N_{0.158}P_{0.009}S_{0.0035}K_{0.0015} + 7.4400\, CO_2 + 8.0963\, H_2O + 0.6133\, H^+ \quad 3$$

$$C_{12}H_{22}O_{11} + 0.0410\, HPO_4^{2-} + 0.0160\, SO_4^{2-} + 0.0068\, K^+ + 7.3790\, O_2 + 0.7205\, NH_4OH \rightarrow 4.5600\, CH_{1.78}O_{0.66}N_{0.158}P_{0.009}S_{0.0035}K_{0.0015} + 7.4400\, CO_2 + 8.7095\, H_2O + 0.1072\, OH^- \quad 4$$

$$C_{12}H_{22}O_{11} + 0.0410\, HPO_4^{2-} + 0.0160\, SO_4^{2-} + 0.0068\, K^+ + 7.9189\, O_2 + 0.7205\, NH_2CONH_2 \rightarrow 4.5600\, CH_{1.78}O_{0.66}N_{0.158}P_{0.009}S_{0.0035}K_{0.0015} + 8.1600\, CO2 + 8.3495\, H_2O + 0.1072\, OH^- \quad 5$$

Protons are only formed during growth with ammonium which explains the observed substantial consumption of KOH by the pH controller with ammonium sulfate; the proton release equates to 0.28 g consumed KOH per g grown biomass and hence somewhat deviates from the measured KOH consumption (0.34 $g_{KOH}/g_X$). With $NH_4OH$ or urea (Equations 4 and 5), the balances predict a slight alkalinization since $OH^-$ ions are formed. The consumption of some KOH with urea as an N source is contradictory to this finding, but it should be kept in mind that the final pH was higher than the initial pH (Figure 3B); synthesis of organic acids (acetate, pyruvate, 2-oxoglutarate, and succinate were by-products of aerobic sugar metabolism of K. marxianus[16],[68],[70],[71]) presumably caused KOH consumption, and consumption of these acidic metabolites after depletion of sugar alkalinized the medium. But such temporary metabolite accumulation was not considered at balancing. Another interfering effect originates from whey-borne lactate (ca. 4 g/L in DW medium [10]) whose microbial utilization also causes some alkalinization.

The balance equations allow to compare calculated with measured masses of consumed oxygen and formed carbon dioxide. The expected masses were calculated from the masses of utilized sugar. The measured masses (Table 4) were 1% to 20% smaller than predicted for unknown reason, but the ratio between formed $CO_2$ and consumed oxygen (the average RQ values) was ca. 1.02 mol/mol (Table 4) and agreed well with the predicted values (1.01 to 1.03 mol/mol).

# CONCLUSIONS

Whey is poor in nitrogen and requires supplementation with an N source for effective production of yeast biomass. Ammonium sulfate, as usually applied for this reason, causes medium acidification by residual sulfate which requires pH control by alkaline substances to avoid growth inhibition. Application of ammonium carbonate instead of ammonium sulfate is not helpful since added $(NH_4)_2CO_3$ elevates the pH to inhibitory levels. K. marxianus DSM 5422 assimilates urea as an alternative N source. Consumption of urea means proton neutrality, medium acidification is minor, and only a little pH corrective is required. Moreover, the use of urea reduces the salt load (less inhibition, diminished environmental impact). Dosage of ammonia by the pH controller is a cost-saving alternative, but a suitable supplement of $(NH_4)_2SO_4$ is needed as a source of sulfur and for initiating dosage of ammonia.

## AUTHORS' CONTRIBUTIONS

CL and TU conceived of the study. EG, CL, and TU explored relevant literature. CL and TU designed the experiments. EG and TU conducted the experiments. CL performed data analysis. CL and TB drafted the manuscript. All authors read and approved the final manuscript.

# ACKNOWLEDGEMENTS

Thanet Urit would like to express his thanks to the Nakhon Sawan Rajabhat University (Muang Nakhon Sawan, Thailand) for financial support. We are grateful to Mrs. E. Kneschke for technical assistance, to M. Heller from the Sachsenmilch Leppersdorf GmbH (Germany) for providing whey permeate, and to A. Stukert for performing two bioreactor experiments.

# REFERENCES

1. Posada JA, Patel AD, Roes A, Blok K, Faaij APC, Patel MK (2013) Potential of bioethanol as a chemical building block for biorefineries: preliminary sustainability assessment of 12 bioethanol-based products. Bioresour Technol 135:490-499

2. Kim S-J, Jung S-M, Park Y-C, Park K (2007) Lipase catalyzed transesterification of soybean oil using ethyl acetate, an alternative acyl acceptor. Biotechnol Bioprocess Eng 12:441-445

3. Modi MK, Reddy JRC, Rao BVSK, Prasad RBN (2007) Lipase-mediated conversion of vegetable oils into biodiesel using ethyl acetate as acyl acceptor. Bioresour Technol 98:1260-1264

4. Uthoff S, Bröker D, Steinbüchel A (2009) Current state and perspectives of producing biodiesel-like compounds by biotechnology. Microb Biotechnol 2:551-565

5. Röttig A, Wenning L, Bröker D, Steinbüchel A (2010) Fatty acid alkyl esters: perspectives for production of alternative biofuels. Appl Microbiol Biotechnol 85:1713-1733

6. Hwang S-CJ, Lee C-M, Lee H-C, Pua HF (2003) Biofiltration of waste gases containing both ethyl acetate and toluene using

different combinations of bacterial cultures. J Biotechnol 105:83-94

7.   Kam S-K, Kang K-H, Lee M-G (2005) Removal characteristics of ethyl acetate and 2-butanol by a biofilter packed with jeju scoria. J Microbiol Biotechnol 15:977-983

8.   Chan W-C, Su M-Q (2008) Biofiltration of ethyl acetate and amyl acetate using a composite bead biofilter. Bioresour Technol 99:8016-8021

9.   Löser C, Urit T, Bley T (2014) Perspectives for the biotechnological production of ethyl acetate by yeasts. Appl Microbiol Biotechnol 98:5397-5415

10.  Urit T, Löser C, Wunderlich M, Bley T (2011) Formation of ethyl acetate by Kluyveromyces marxianus on whey: studies of the ester stripping. Bioprocess Biosyst Eng 34:547-559

11.  Urit T, Löser C, Stukert A, Bley T (2012) Formation of ethyl acetate by Kluyveromyces marxianus on whey during aerobic batch cultivation at specific trace-element limitation. Appl Microbiol Biotechnol 96:1313-1323

12.  Urit T, Manthey R, Bley T, Löser C (2013) Formation of ethyl acetate by Kluyveromyces marxianus on whey: influence of aeration and inhibition of yeast growth by ethyl acetate. Eng Life Sci 13:247-260

13.  Löser C, Urit T, Stukert A, Bley T (2013) Formation of ethyl acetate from whey byKluyveromyces marxianus on a pilot scale. J Biotechnol 163:17-23

14.  Urit T, Li M, Bley T, Löser C (2013) Growth of Kluyveromyces marxianus and formation of ethyl acetate depending on temperature. Appl Microbiol Biotechnol 97:10359-10371

15.  Löser C, Urit T, Nehl F, Bley T (2011) Screening of Kluyveromyces strains for the production of ethyl acetate: design and evaluation of a cultivation system. Eng Life Sci 11:369-381

16.  Löser C, Urit T, Förster S, Stukert A, Bley T (2012) Formation of ethyl acetate byKluyveromyces marxianus on whey during aerobic batch and chemostat cultivation at iron limitation. Appl Microbiol Biotechnol 96:685-696

17.  Kallel-Mhiri H, Engasser J-M, Miclo A (1993) Continuous ethyl acetate production byKluyveromyces fragilis on whey permeate. Appl Microbiol Biotechnol 40:201-205

18. Willetts A (1989) Ester formation from ethanol by Candida pseudotropicalis. Antonie Van Leeuwenhoek 56:175-180

19. Aziz S, Memon HUR, Shah FA, Rajoka MI, Soomro SA (2009) Production of ethanol by indigenous wild and mutant strain of thermotolerant Kluyveromyces marxianus under optimized fermentation conditions. Pak J Anal Environ Chem 10(1+2):25-33

20. Lertwattanasakul N, Rodrussamee N, Suprayogi LS, Thanonkeo P, Kosaka T, Yamada M (2011) Utilization capability of sucrose, raffinose and inulin and its less-sensitiveness to glucose repression in thermotolerant yeast Kluyveromyces marxianus DMKU 3–1042. AMB Express 1:20 BioMed Central Full Text

21. Fonseca GG, de Carvalho NMB, Gompert AK (2013) Growth of the yeast Kluyveromyces marxianus CBS 6556 on different sugar combinations as sole carbon and energy source. Appl Microbiol Biotechnol 97:5055-5067

22. Löser C, Urit T, Keil P, Bley T (2014) Studies on the mechanism of synthesis of ethyl acetate in Kluyveromyces marxianus DSM 5422. Appl Microbiol Biotechnol. Accepted, doi:10.1007/s00253-014-6098-4

23. Armstrong DW, Yamazaki H (1984) Effect of iron and EDTA on ethyl acetate accumulation inCandida utilis. Biotechnol Lett 6:819-824

24. Fonseca GG, Heinzle E, Wittmann C, Gombert AK (2008) The yeast Kluyveromyces marxianusand its biotechnological potential. Appl Microbiol Biotechnol 79:339-354

25. Ugalde OU, Castrillo IJ (2002) Single cell proteins from fungi and yeasts. In: Dilip KA, George GK (eds) Applied mycology and biotechnology, Elsevier, Amsterdam. pp 123-149

26. Ghaly AE, Kamal MA (2004) Submerged yeast fermentation of acid cheese whey for protein production and pollution potential reduction. Water Res 38:631-644

27. Schultz N, Chang L, Hauck A, Reuss M, Syldatk C (2006) Microbial production of single-cell protein from deproteinized whey concentrates. Appl Microbiol Biotechnol 69:515-520

28. Yadav JSS, Bezawada J, Elharche S, Yan S, Tyagi RD, Surampalli RY (2014) Simultaneous single-cell protein production and COD removal with characterization of residual protein and

intermediate metabolites during whey fermentation by K. marxianus. Bioprocess Biosyst Eng 37:1017-1029

29. Parrondo J, García LA, Díaz M (2009) Nutrient balance and metabolic analysis in aKluyveromyces marxianus fermentation with lactose-added whey. Brazil J Chem Eng 26:445-456

30. De Nicola R, Hazelwood LA, De Hulster EAF, Walsh MC, Knijnenburg TA, Reinders MJT, Walker GM, Pronk JT, Daran J-M, Daran-Lapujade P (2007) Physiological and transcriptional responses of Saccharomyces cerevisiae to zinc limitation in chemostat cultures. Appl Environ Microbiol 73:7680-7692

31. Mahmoud MM, Kosikowski FV (1982) Alcohol and single cell protein production byKluyveromyces in concentrated whey permeates with reduced ash. J Dairy Sci 65:2082-2087

32. Kar T, Misra AK (1998) Effect of fortification of concentrated whey on growth of Kluyveromyces sp. Rev Argent Microbiol 30(4):163-169

33. Belem MAF, Lee BH (1999) Fed-batch fermentation to produce oligonucleotides fromKluyveromyces marxianus grown on whey. Process Biochem 34:501-509

34. Domingues L, Lima N, Teixeira JA (2001) Alcohol production from cheese whey permeate using genetically modified flocculent yeast cells. Biotechnol Bioeng 72:507-514

35. Moeini H, Nahvi I, Tavassoli M (2004) Improvement of SCP production and BOD removal of whey with mixed yeast culture. Electronic J Biotechnol 7:249-255

36. Aktaş N, Boyacı İH, Mutlu M, Tanyolaç A (2006) Optimization of lactose utilization in deproteinated whey by Kluyveromyces marxianus using response surface methodology (RSM). Bioresour Technol 97:2252-2259

37. Gupte AM, Nair AS (2010) β-galactosidase production and ethanol fermentation from whey using Kluyveromyces marxianus NCIM 3551. J Sci Ind Res 69:855-859

38. Castrillo JI, de Miguel I, Ugalde UO (1995) Proton production and consumption pathways in yeast metabolism. A chemostat culture analysis. Yeast 11:1353-1365

39. Hensing MCM, Bangma KA, Raamsdonk LM, de Hulster E, van Dijken JP, Pronk JT (1995) Effects of cultivation conditions on the

production of heterologous β-galactosidase byKluyveromyces lactis. Appl Microbiol Biotechnol 43:58-64

40. Vicente A, Castrillo JI, Teixeira JA, Ugalde U (1998) On-line estimation of biomass through pH control analysis in aerobic yeast fermentation systems. Biotechnol Bioeng 58:445-450

41. Tovar-Castro L, García-Garibay M, Saucedo-Castañeda G (2008) Lactase production by solid-state cultivation of Kluyveromyces marxianus CDBBL 278 on an inert support: effect of inoculum, buffer, and nitrogen source. Appl Biochem Biotechnol 151:610-617

42. Vivier D, Ratomahenina R, Moulin G, Galzy P (1993) Study of physicochemical factors limiting the growth of Kluyveromyces marxianus. J Ind Microbiol 11:157-161

43. Antoce O-A, Antoce V, Takahashi K (1997) Calorimetric study of yeast growth and its inhibition by added ethanol at various pHs and temperatures. Netsu Sokutei 24(4):206-213

44. Large PJ (1986) Degradation of organic nitrogen compounds by yeasts. Yeast 2:1-34

45. Castrillo JI, Ugalde UO (1992) Energy metabolism of Kluyveromyces marxianus in deproteinated whey. Chemostat studies. Modelling. J Biotechnol 22:145-152

46. Castrillo JI, Ugalde UO (1993) Patterns of energy metabolism and growth kinetics ofKluyveromyces marxianus in whey chemostat culture. Appl Microbiol Biotechnol 40:386-393

47. Rech R, Cassini CF, Secchi A, Ayub MAZ (1999) Utilization of protein-hydrolyzed cheese whey for production of β-galactosidase by Kluyveromyces marxianus. J Ind Microbiol Biotechnol 23:91-96

48. Hortsch R, Löser C, Bley T (2008) A two-stage CSTR cascade for studying the effect of inhibitory and toxic substances in bioprocesses. Eng Life Sci 8:650-657

49. Rahmatullah M, Boyde TRC (1980) Improvements in the determination of urea using diacetyl monoxime; methods with and without deproteinisation. Clin Chim Acta 107:3-9

50. Francis PS (2006) The determination of urea in wine - a review. Aust J Grape Wine Res 12:97-106

51.  Smithers GW (2008) Whey and whey proteins - from 'gutter-to-gold'. Int Dairy J 18:695-704

52.  Prazeres AR, Carvalho F, Rivas J (2012) Cheese whey management: a review. J Environ Manage 110:48-68

53.  Cordier J-L, Butsch BM, Birou B, von Stockar U (1987) The relationship between elemental composition and heat of combustion of microbial biomass. Appl Microbiol Biotechnol 25:305-312

54.  Krzystek L, Ledakowicz S (2000) Stoichiometric analysis of Kluyveromyces fragilis growth on lactose. J Chem Technol Biotechnol 75:1110-1118

55.  Silva-Santisteban BOY, Converti A, Filho FM (2006) Intrinsic activity of inulinase fromKluyveromyces marxianus ATCC 16045 and carbon and nitrogen balances. Food Technol Biotechnol 44:479-483

56.  Hamme V, Sannier F, Piot J-M, Didelot S, Bordenave-Juchereau S (2009) Crude goat whey fermentation by Kluyveromyces marxianus and Lactobacillus rhamnosus: contribution to proteolysis and ACE inhibitory activity. J Dairy Res 76:152-157

57.  Foukis A, Stergiou P-Y, Theodorou LG, Papagianni M, Papamichael EM (2012) Purification, kinetic characterization and properties of a novel thermo-tolerant extracellular protease fromKluyveromyces marxianus IFO 0288 with potential biotechnological interest. Bioresour Technol 123:214-220

58.  Fox PF (2003) Significance of indigenous enzymes in milk and dairy products. In: Whitaker JR, Voragen AGJ, Wong DWS (eds) Handbook of food enzymology, Marcel Dekker Inc, New York, Basel. pp 255-277

59.  Perea A, Ugalde U, Rodriguez I, Serra JL (1993) Preparation and characterization of whey protein hydrolysates: applications in industrial whey bioconversion processes. Enzyme Microb Technol 15:418-423

60.  Walker GM, White NA (2005) Introduction to fungal physiology. In: Kavanagh K (ed) Fungi: biology and applications, John Wiley & Sons Ltd, Chichester. pp 1-34

61.  Nahvi I, Moeini H (2004) Isolation and identification of yeast strains with high beta-galactosidase activity from dairy products. Biotechnol 3(1):35-40

62. Rajoka MI, Khan S, Latif F, Shahid R (2004) Influence of carbon and nitrogen sources and temperature on hyperproduction of a thermotolerant β-glucosidase from synthetic medium byKluyveromyces marxianus. Appl Biochem Biotechnol 117(2):75-92

63. Whitney PA, CooperTG (1972) Urea carboxylase and allophanate hydrolase. Two components of adenosine triphosphate: urea amido-lyase in Saccharomyces cerevisiae. J Biol Chem 247:1349-1353

64. Shay LK, Hunt HR, Wegner GH (1987) High-productivity fermentation process for cultivating industrial microorganisms. J Ind Microbiol 2:79-85

65. Hack CJ, Marchant R (1998) Characterisation of a novel thermotolerant yeast, Kluyveromyces marxianus var marxianus: development of an ethanol fermentation process. J Ind Microbiol Biotechnol 20:323-327

66. Nor ZM, Tamer MI, Scharer JM, Moo-Young M, Jervis EJ (2001) Automated fed-batch culture ofKluyveromyces fragilis based on a novel method for on-line estimation of cell specific growth rate. Biochem Eng J 9:221-231

67. Gélinas P, Barrette J (2007) Protein enrichment of potato processing waste through yeast fermentation. Bioresour Technol 98:1138-1143

68. Fonseca GG, Gombert AK, Heinzle E, Wittmann C (2007) Physiology of the yeastKluyveromyces marxianus during batch and chemostat cultures with glucose as the sole carbon source. FEMS Yeast Res 7:422-435

69. Mazutti MA, Zabot G, Boni G, Skovronski A, de Oliveira D, Di Luccio M, Rodrigues MI, Maugeri F, Treichel H (2010) Mathematical modeling of Kluyveromyces marxianus growth in solid-state fermentation using a packed-bed bioreactor. J Ind Microbiol Biotechnol 37:391-400

70. Silva-Santisteban BOY, Converti A, Filho FM (2009) Effects of carbon and nitrogen sources and oxygenation on the production of inulinase by Kluyveromyces marxianus. Appl Biochem Biotechnol 152:249-261

71.  Rocha SN, Abrahão-Neto J, Gombert AK (2011) Physiological diversity within theKluyveromyces marxianus species. Antonie Van Leeuwenhoek 100:619-630

# Genetic Improvement of Microorganisms for Applications in Biorefineries

Bárbara G Paes[1,2] and João RM Almeida[2]

[1]Cellular Biology Department Institute of Biological Sciences University of Brasília, Brasilia, 70910-900, DF, Brazil

[2]Embrapa Agroenergy, Parque Estação Biológica W3 Norte, Brasília, 70770-901, DF, Brazil

## ABSTRACT

The development of biorefineries directed to the production of fuels, chemicals and energy is important to reduce economic dependence and environmental impacts of a petroleum-based economy. Microorganisms are essential in several industrial bioprocesses nowadays, and it is expected that new microbial bioprocesses will play a key role in biorefineries. However, the bioconversion process requires a robust and highly productive microorganism. In this scenario, several strategies to genetically improve microorganisms to overcome the bioprocesses challenges have been considered. In this work, we review microorganisms importance in the biorefineries concept, highlight the desirable traits they must hold in order to be employed, and discuss the main strategies to improve such traits. The focuses of this work are on four main targets in the improvement of microorganisms: driving carbon flux towards the desired pathway, increasing tolerance to toxic compounds, increasing substrate uptake range and new products generation.

# INTRODUCTION

The interest in renewable and sustainable biotechnological processes for energy, biofuels and chemicals production has been increasing over the years. Economical and environmental factors have been pushing the chemical industry, for instance, to invest in new means to get the same products in a more sustainable and economical way. It is estimated that by 2025, 15% of global chemical sales will be bio-derived [1]. In this context, the development of biorefineries appears as an important alternative to the common known petroleum-based processes and products. Biorefineries can be defined as "the sustainable processing of biomass into a spectrum of marketable products (food, feed, material, chemicals) and energy (fuels, power, heat) [2]. Chemical, physical and biological processes can be employed in a biorefinery to convert biomass into a large spectrum of products of interest [2]. The biorefinery concept is attractive because it would allow production of high added-value compounds and/or big volumes of biofuels, with market competitive prices, while reducing waste disposal and energy costs. In addition, always taking into consideration the sustainability of the process and its indirect impacts (such as water use, impact on soil and biodiversity and competition for food). Few biorefineries have been in operation for several years, for instance, the pulp and paper based biorefinery Borregaard, in Norway [3], but there is still immense potential to be developed in several countries [2].

Microorganisms are main characters in industrial bioprocesses, being directly responsible for the production of the desired chemical, or indirectly providing important components for the processes. Indeed, there are several industrial processes in operation around the world based on microorganisms for production of food additives, enzymes and chemicals; for instance, the bioethanol industry from sugarcane in Brazil. Due to their versatility, microorganisms are also expected to play an essential role in conversion processes employed in biorefineries.

The development of biorefineries brings new opportunities and challenges to the industrial application of microorganisms. New substrates may be used and a variety of products formed; however, strains adapted to the industrial processes need to be developed. Based on the feedstock characteristics and desired products, a microbial strain and a production process optimization are needed to achieve an ideal conversion process (Figure 1). This led research groups from several institutions worldwide to start developing microbial strains for application in biorefineries and nowadays there are many promising experimental processes being developed. This work focuses on the importance of microorganisms in industries, summarizes and discusses the main targets for microbial improvement and the strategies currently employed to generate and improve strains to achieve commercial, technological and environmentally viable industrial processes (Figure 1). In addition, we highlight and exemplify general strategies to develop microorganisms that are able to produce fuels and chemicals from renewable feedstocks.

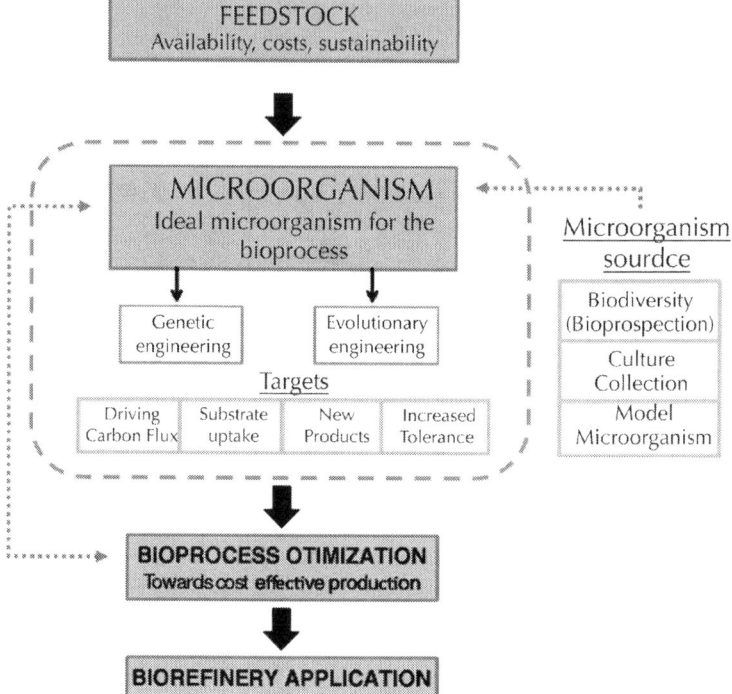

**Figure 1:** Main steps for the development of a new bioprocess integrated to a biorefinery (green boxes). Strategies to improve microbial traits (orange lined boxes) and most common targets for microorganisms' improvement (green lined boxes) are also shown. The importance of having an improved microorganism and an optimized bioprocess in which it should be applied is also indicated (gray arrow in the left).

# REVIEW

## Microorganisms Currently Used and New from Biodiversity

In order for a microorganism to be applied in a bioprocess, it must present specific traits, which would allow its maximum performance, i.e. high production yields and rates, even when submitted to one or a series of challenges. These may include substrate and product toxicity, variations or extreme pH values, high or variable temperature and pressure values, presence of competitors (biological or chemical contamination), inability to use all components of the substrate, and others[4]. It will be rather difficult to find a microorganism that has naturally all necessary traits to be employed in an industrial bioprocess. Therefore, genetic improvements of microorganisms have become an essential step in the development of such processes (Figure 1).

Genetically, physiologically and biologically well-characterized microorganisms, such as the yeast *Saccharomyces cerevisiae*, the bacteria *Escherichia coli* and other microorganisms which are also already employed commercially, are frequently the initial choice for the development of novel biocatalysts for industrial application. Previous knowledge about such microbes would ease the task of genetic improvement and the industrial utilization of new strains.

The yeast *S. cerevisiae* is the eukaryotic model microorganism and it is commonly utilized in bioethanol, brewery and bakers industries worldwide. In addition, several new bioprocesses for fuels and chemicals production are being developed based on this yeast [4] (Figure 2). It presents high level of ethanol tolerance (i.e. product inhibition is absent or minimal), ability to grow under different aeration conditions, including strictly anaerobic (which makes the process more easily controlled), have little nutrition requirements, shows high tolerance to toxic compounds and low pH tolerance, which also contributes to prevent bacterial contamination [5]. The accumulated knowledge about *S. cerevisiae* surpasses any other eukaryotic species. In fact, *S. cerevisiae*'s genome was the first eukaryotic genome to be completely sequenced [6]. More recently, systems biology tools are abundantly available and further studies have been improving the understanding about this organism. Indeed, *S. cerevisiae* has been genetically modified to produce a variety of chemical products (Table 1 and Figure 2).

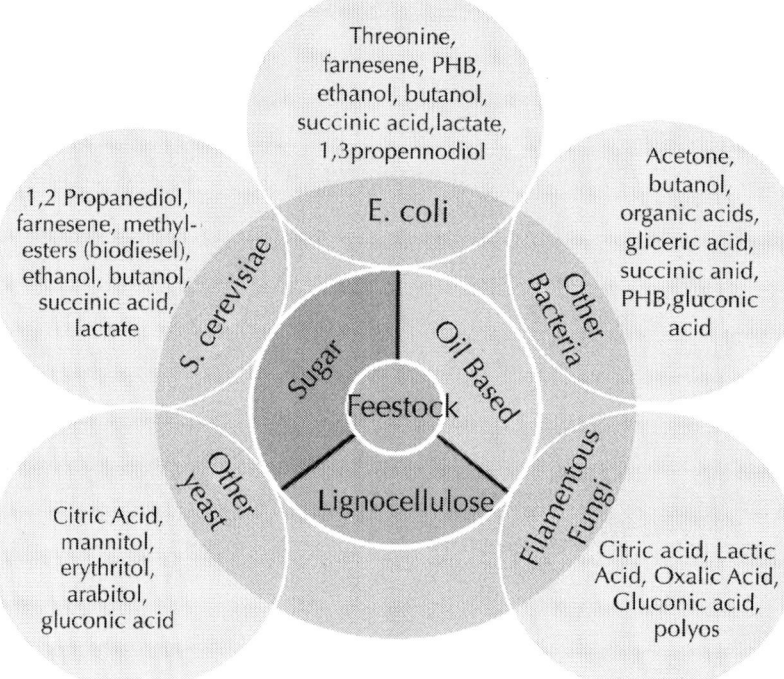

**Figure 2:** Examples of biofuels and chemicals produced from renewable feedstock by microbial strains. Detailed information about microbial strains and their production capacity can be found at 3, 8, 21, 29, 32, 34 and references there in.

**Table 1:** Selected examples of four major target categories where genetic engineering strategies were applied to improve product formation by micro-organisms

| Organism | Product | Main sub-strate | Yield* | Productiv-ity | Concentra-tion | Outcomes | Main genetic modifica-tions | Refer-ence |
|---|---|---|---|---|---|---|---|---|
| Driving carbon flux towards the desired pathway | | | | | | | | |
| E. coli SY4 | Ethanol | Glycerol | 0.42 g g⁻¹ | 0.15 g L⁻¹ h⁻¹ | 7.8 g L⁻¹ | Yield improved 69 fold. Engi-neered strains efficiently uti-lized glycerol in a minimal medium without rich supplements | Deletion of genes to minimize the synthesis of by-products | [7] |
| E. coli LA02Δdld | Lactic acid | Glycerol | 0.80 g g⁻¹ | 1.25 g g⁻¹ h⁻¹ | 32 g L⁻¹ | Low-value glycerol streams to a higher- value product like D-lactate. Yield im-proved seven fold | Overex-pression of pathways involved in the conver-sion of glyc-erol to lactic acid and blocking those lead-ing to the synthesis of competing by-products | [8] |
| E. coli | Acetate | Glucose | 0.456 g g⁻¹ | 1.38 g g⁻¹ h⁻¹ | 53 g L⁻¹ | Reduc-tion of the fermentation by products concentra-tion by 1, 25 (succinate) to 33 fold (lactate). Yield improved over seven fold | Deletion of genes in-volved in the succinate formation as fermentation product | [9] |

| | | | | | | | | |
|---|---|---|---|---|---|---|---|---|
| *Y. lipolytica* | Succinic acid | Glycerol | 0.45 g g$^{-1}$ | n.d | 45 g L$^{-1}$ | Succinic acid production yield increased over 20 fold | Deletion in the gene coding one of succinate dehydrogenase subunits | [10] |
| *Y-3314* | | | | | | | | |
| *Mannheimia succiniciproducens* | Succinic Acid | Glucose | 0.76 g g$^{-1}$ | 1.8 g g$^{-1}$ h$^{-1}$ | 52.4 g L$^{-1}$ | Nearly complete elimination of fermentation by-products, (acetic, formic, and lactic acids) and carbon recovery increased to 58% to 77% by fed-batch culture | Disruption of genes responsible for by product formation (*ldhA, pflB, pta,* and *ackA* ) | [11] |
| **Increasing of tolerance to toxic compounds** | | | | | | | | |
| *C. acetobutylicum* | Butanol | Glucose | n.d. | n.d. | | Increased tolerance and extended metabolism response to butanol stress. | Overexpression of spo0A, responsible for the transcription of solvent formation genes | [12] |
| *C. acetobutylicum* | Butanol | Glucose | 70.8% | n.d. | 13.6 g L$^{-1}$ | Reduction of acetone production from 2,83 g L$^{-1}$ to 0,21 g L$^{-1}$ and enhanced butanol yield from 57% to 70.8 | Disruption of the acetoacetate decarboxylase gene (*adc*) avoiding acetone production and optimization of medium | [13] |

| S. cerevisiae | Ethanol | Glucose plus HMF (inhibitor) | 0.43 g g⁻¹ | 0.61 g g⁻¹ h⁻¹ | n.d | Four times higher specific uptake rate of HMF and 20% higher specific Ethanol productivity | Overexpression of alcohol dehydrogenases ADH6 or ADH1-mutated | [14] |
|---|---|---|---|---|---|---|---|---|
| S. cerevisiae | Ethanol | Spruce hydrolystae | n.d | 0.39 g g⁻¹ h⁻¹ | n.d | HMF conversion rate and ethanol productivity for the engineered strains four to five times and 25% higher than for the control strain. | Overexpression of alcohol dehydrogenases ADH6 or ADH1-mutated | [14] |
| E. coli XW068(pLOI4319) | Lactate | Xylose plus HMF | 85% of the theoretical maximum | n.d. | n.d | Furfural tolerance increased by 50%. Minimal growth and lactate production occurred after 120 h for the control strain | Overexpression of NADH-dependent propanediol oxidoreductase (FucO) | [15] |
| **Increasing substrate uptake range** | | | | | | | | |
| E. coli | Ethanol | Xylose | 0.48 g g⁻¹ | 2.00 g g⁻¹ h⁻¹ | 43 g L⁻¹ | Rapid co-fermentation due to reduced repression of xylose metabolism by glucose, and 60% less time required for fermentation of 5-sugars mix to ethanol. | Deletion of methylglyoxal synthase gene (mgsA), involved in sugar metabolism | [16] |

| Lactobacillus plantarum | Lactic Acid | Corn starch | 0.89 g g$^{-1}$ | 4.51 g g$^{-1}$ h$^{-1}$ | 86 g L$^{-1}$ | First direct and efficient fermentation of optically pure D- lactic acid from raw corn starch reported | Deletion of L-lactate dehydrogenase gene (ldhL1) and expression of *Streptococcus bovis* 148 α-amylase (AmyA) | [17] |
|---|---|---|---|---|---|---|---|---|
| *S. cerevisiae* | Ethanol | Xylose | 0.43 g g$^{-1}$ | 0.02 g g$^{-1}$ h$^{-1}$ | 7.3 g L$^{-1}$ | Higher ethanol yields than XR/ XDH carrying strains | Overexpression of *Piromyces sp* xylose isomerase (XI) | [18] |
| *S. cerevisiae* | Ethanol | Xylose | 0.33 g g$^{-1}$ | 0.04 g g$^{-1}$ h$^{-1}$ | 13.3 g L$^{-1}$ | Higher specific ethanol productivity and final ethanol concentration than XI carrying strains | Overexpression of xylose reductase (XR) and xylitol dehydrogenase (XDH) enzymes from *Scheffersomyces stipitis* | [19] |
| *E. coli* | Butanol | Glucose | 6.1% | 0.02 g g$^{-1}$ h$^{-1}$ | 1.2 g L$^{-1}$ | Anaerobic production of butanol by a microorganism expressing genes from a strict aerobic organism | Expression of *C. acetobutylicum* butanol pathway sinthetic genes in *E. coli* | [20] |
| **Generation of new products** | | | | | | | | |
| *E. coli* | Fatty acid ethyl esters (FAEEs) | Glucose | 7% | n.d. | 30.7 g L$^{-1}$ | Tailored fatty ester (biodiesel) production | Heterologous expression of a "FAEE pathway" engineered in *E. coli* | [21] |

| S. cerevisiae | Butanol | Galactose | n.d | n.d | 2.5 mg L$^{-1}$ | First demonstration of n-butanol production in S. cerevisiae | N-butanol biosynthetic pathway engineered in S. cerevisiae | [22] |
| E. coli K12 | 1,3-propandiol | Glycerol | 90.2% | 2.61 g g$^{-1}$ h$^{-1}$ | 104.4 g L$^{-1}$ | Substantially high yield and productivity efficiency of 1,3-PD with glycerol as the sole source of carbon | Heterologous over-expression of genes from natural producers of 1,3-PDO | [23]. |

*expressed in g product per g substrate or % of maximum theoretical; n.d. not determined; n.c. not calculated.

Paes and Almeida

Paes and Almeida *Chemical and Biological Technologies in Agriculture* 2014 1:21   doi:10.1186/s40538-014-0021-1

While *S. cerevisiae* is the eukaryotic model organism, *E. coli* stands as the prokaryotic one. Similarly to the yeast, its genetics, physiology and biology are well-known, and genetic manipulation tools are already well established for it. *E. coli*'s complete genome was published in 1997 [24]. *E. coli* has been genetically modified to produce many different chemical compounds (Table 1 and Figure 2) [25] demonstrating its biotechnological potential. Having many characteristics of interest to industry, such as efficient growth under industrial conditions, low nutritional requirements, anaerobic growth, capacity to use many different carbon sources including carbohydrates, polyols, and fatty acids [26] , this bacterium was already engineered to produce ethanol from lignocellulose [27] and it is currently employed in a bioethanol pilot plant in Florida [28].

The accumulated knowledge about such "model" microorganisms, in the most different subjects of studies, and familiarization with their requirements and performance, facilitates the task of genetic improvement and eases the industrial utilization of new strains. However, identification of new microorganisms, new genes and enzymes from the microbial biodiversity, still remains essential to reveal new traits and capabilities to favor development of biotechnological applications [29]. This becomes more relevant because estimates point that there are 10 $^{30}$ microorganism cells in the Earth. That is more than stars in the universe. And like in the universe, we only know a small fraction of those, and have characterized even fewer. Indeed, the vast microbial diversity in microorganism collections world-wide still remains unexploited and in the wild unknown species relies an enormous unknown potential.

The discovery of new genes, pathways, enzymes and characteristics in newly discovered and described wild organisms, can be applied in the development of new production processes. For instance, *Pichia ciferrii* was recognized as a potential producer of sphingolipids, including sphingosine, since genes encoding enzymes of the biosynthetic pathway were identified. However, no detectable amounts of sphingosine were produced by the wild type strain. Thus, metabolic engineering strategies including the implementation and improvement of a metabolic pathway for the conversion of sphinganine to sphingosine were used to develop a final strain capable of producing approximately 240 mg.L$^{-1}$ triacetylated sphingosine (TriASo) in shake flasks and up to 890 mg.kg$^{-1}$ in lab-scale fermentation. Further improvement of such strain could lead to even higher concentrations of sphinganine and sphingosine for cosmetic and pharmaceutical applications[30]. These results are still preliminary for industrial application, but they clearly demonstrate the potential of bioprospecting for developing bioprocesses.

Most certainly, encountering the complete ideal wild microorganism to be used in a specific biorefinery is a challenging mission. Scarcely a wild microorganism will have all desired traits to be employed in a biorefinery. Thus, genetic engineering strategies shall be used to design an ideal host, improving substrate uptake range and product formation, increasing tolerance, yields and rates and allowing production of new chemicals by a specific strain. In the next sections, strategies to develop such microorganisms for industrial processes applications are presented and discussed.

# Genetic Improvement of Microorganisms

Bioprocesses require microbial strains that are able to tolerate several different stresses while keeping high yields and productivity. In addition, in order to develop and keep viable bioprocesses, the microbial strains employed or envisioned to be used need constant genetic improvement for achieving or keeping high production rates. For instance, even though the yeast*S. cerevisiae* is used for more than 30 years in Brazilian bioethanol industry, each year improved strains for the process are selected [31]. On the other hand, wild strains that contain desirable characteristics for biotechnological application usually have very low production rates or are very sensitive to the industrial conditions. Thus, different strategies have been applied to genetically improve microorganisms to solve problems such as the ones listed above, and directly or indirectly increase productivity and consequently the profitability of the bioprocess. Four major target categories where genetic and evolutionary engineering strategies may be applied to improve product formation by microorganisms are: i) driving carbon flux, ii) increase tolerance to toxic compounds, iii) increase of substrate uptake range and iv) generation of new products (Figure 1). Following, each of these targets is discussed.

# Driving Carbon Flux

Naturally, microorganisms have their metabolic pathway optimized to sustain maximal growth and outcome competitors in the environment. Thus, production of a desired chemical usually is reduced during cell growth (expenses of carbon and energy sources) and by-product formation. Thus, a common target for modifications that directly affects microorganism's productivity is driving carbon flux through a specific pathway towards the desired product.

Microorganisms from the most different groups, from bacteria and yeast to filamentous fungi, have been genetically modified to increase production of a desired biofuel or chemical compound. Nowadays, strains that are able to produce a variety of chemical compounds in concentrations as high as above 90% m/m of the theoretical maximum are available (Table 1) (Figure 2). The strategies to increase product formation generally include a series of modifications in the microorganism metabolism, achieved by overexpression or knockout of enzymes in the producing pathway [13],[32], changing redox balancing of the cell by redirecting carbon fluxes from NADPH- to NADH consuming reactions [33]-[36], engineering global transcription machinery [37] and others (Table 1). All these types of modification were employed, for instance, to obtain *S. cerevisiae* strains that are able to produce ethanol from sugars that are present in lignocellulosic hydrolysates [38]. *S cerevisiae* strains able to ferment lignocellulosic hydrolysates rich in xylose and produce ethanol with yields up to 0.44 g ethanol/g sugar (86% of theoretical maximum) were obtained [38].

# Increased Tolerance to the Substrate

Another common trait that may hamper product formation by microbial strains is their low tolerance to substrate or fermentation end-product. Indeed, the fermentation medium may impose a harsh environment for the microorganism and consequently, an important trait to define the strain to be used in an industrial process is its tolerance level to toxic compounds. When tolerant strains are not available for the desired process, genetic engineering strategies may be applied to improve strain response for inhibitory compounds. A good example of such is the improvement of strains for production of biofuels and chemicals from lignocellulosic hydrolysates.

Lignocelulose is composed of the polymers cellulose and hemicellulose, and the macromolecule lignin. Prior to fermentation, lignocellulosic biomass must be submitted into a pretreatment to reduce its recalcitrance. In the next step, the hydrolysis, cellulose and hemicellulose are broken down into their sugar monomers, those which should later be converted into the final product [38]. The problem is that during pretreatment and hydrolysis not only sugars are solubilized, but also, compounds that inhibit microbial metabolism may be released and formed during these steps [39]. Indeed, compounds like furaldehydes (5-hydroxymethyl-2-furaldehyde – HMF; - and 2-furaldehyde – furfural), organic acids (acetic, levulinic and furoic) and phenolic derivatives are commonly found in lignocellulosic hydrolysates. However, concentration of such compounds varies according to biomass and process conditions employed. As these inhibitors can affect microbial growth, decrease product yield and productivity; prolong lag phase of microbial growth, and reduce cellular viability [39],[40], several evolutionary or metabolic engineering strategies have been employed to develop strains able to tolerate them. Evolutionary engineering mimics the evolutionary mechanisms of nature, in which through variation, strains are selected according to the response to the pressure they are submitted to [40].

Evolutionary engineering strategies have been applied, for example, to generate strains with higher tolerance to specific compounds (furfural, for instance), or to lignocellulosic hydrolysates, by selecting strains with the ability to remain viable and keep growth even in presence of such compounds. Through multiple selection cycles, in presence of increasing concentrations of the selection pressure, i.e. the toxic compound, mutants with higher tolerance can be selected. To increase genetic variation in the population to be submitted to the selection, pressure mutagenic agents like UV light and EMS (Ethyl methanesulfonate) can be applied. Evolutionary engineering strategies have been commonly employed to obtain *S. cerevisiae*[41], *P. stipitis*[42][43], *S. passalidarum*[44] mutants which are able to ferment lignocellulosic hydrolysates with higher rates than the native strains. For instance, the yeast strain TMB3400 was grown in minimal medium containing 3 mM furfural. Once cells reached late exponential phase they were transferred to a fresh media amended with furfural. Upon shorter lag phases the furfural concentration was increased continuously. Finally, after approximately 300 generations, single colonies were obtained, and the best isolated strain showed a lag phase of 17 h instead of 90 h for parental strain in media supplemented with 17 mM furfural. In addition, viability tests in furfural containing medium showed that the evolved strain remained viable, whereas the parental strain showed continuously decreasing colony- forming unit capacity after 10 h [41]. The main disadvantage of evolutionary engineering resides in the fact that the genetic trait responsible for the improvement has to be identified posteriorly and thus cannot be directly transferred to another strain.

Yeast tolerance to lignocellulosic hydrolysate inhibitors has also been improved by genetic engineering strategies (Table 1) [40]. The general strategy involves identification of genes that confer resistance to inhibitors and their posterior overexpression in the desired microorganism.Yeast oxide-reductases enzymes, like alcohol dehydrogenase 6 (Adh6), and Adh1, able to convert HMF and furfural to their corresponding alcohols, when overexpressed have been shown to improve yeast growth and fermentation rates not only in medium supplemented with theses inhibitors but also in lignocellulosic hydrolysates. Genes related to regeneration of cofactors NAD(P)H, and transcription factors related to stress response have also been demonstrated to increase yeast tolerance towards lignocellulosic hydrolysate inhibitors [45]. Despite the time frame required to identify genes or enzymes that confer increased tolerance, the genetic engineering strategies are advantageous because the trait can be transferred from one strain to another promptly.

# Increase of Substrate Uptake Range

The increased interest to produce fuels and chemicals from renewable resources, especially from lignocellulosic feedstocks and crude glycerol residue from biodiesel industry, made the expansion of substrate utilization another important target for genetic improvement of microorganisms [7]. Screening and genetic engineering of wild- and well-known microbial strains to increase production of fuels an chemicals from substrates previously not- or poorly utilized have gained much attention lately [2],[46].

A better utilization of lignocellulosic feedstocks for fuels and chemicals production requires xylose utilization. This pentose sugar is present in several biomasses and it is the second most abundant sugar in many of them. In sugar cane bagasse, for instance, xylose corresponds to up to 30% of the sugars present in the biomass [47]. Thus, xylose utilization in biotechnological processes is desirable and might contribute considerably to the economic viability of the process. In this context, second generation bioethanol production from xylose with *S. cerevisiae* is one of the most evaluated bioprocess. As this yeast is widely used in alcohol industries, including first generation bioethanol production in Brazil, but it is not able to ferment pentoses, many strategies to construct xylose-fermenting *S. cerevisiae* strains have been employed. Among these, introduction and improvement of xylose catabolic pathways; increase sugar uptake rate by overexpression of transporters; changing redox metabolism; and others as reviewed by Van Maris et al. [5] and Hahn-Hägerdal et al. [38]. Nowadays, several yeast strains able to convert xylose to ethanol are available, either with reductase -dehydrogenase or xylose isomerase pathway, with yields around 90% from the theoretical maximum, (Table 1) [48].

Production of biodiesel by (trans)esterification of oils and fats results in approximately one ton of crude-glycerol from every ten tons of biodiesel produced. As biodiesel production increased worldwide, glycerol availability did too and its prices in the market decreased. Thus, microbial processes to convert glycerol into renewable fuels and chemicals have been considered. Indeed, several groups demonstrated the potential of bacteria utilization , as well as yeast and filamentous fungi for production of ethanol, butanol, 1,3-propanediol, polyols and other chemicals from glycerol (Table 1) (Figure 2). This subject has been recently reviewed by Almeida et al [46] and Yang [49].

# New Products

In addition to increased production rates by redirecting carbon fluxes, increase of substrate uptake ranges and improving tolerance to inhibitory compounds, genetic engineering strategies can be employed to generate microorganisms able to produce biofuels and chemicals not naturally formed by their genetic and biochemical machinery. In this case, enzymes and pathways from one organism can be transferred to the desired microbial host, which ultimately will produce the desired compound. Nowadays, there are several examples of engineered microorganisms for production of compounds such as building block chemicals (compounds from which a big number of molecules of interest can be obtained) rather than bioethanol in this category (Table 1).

Acids derived from lignocellulosic sugars have a large potential as precursors of plastics and as building block compounds [49]. Among these there is xylonic acid, an organic acid with five carbons, derived of xylose, which is naturally produced by bacteria from the genre *Acetobacter,Aerobacter, Pseudomonas Gluconobacter* and *Erwinia*. Although wild type bacteria are efficient in the xylonic acid production, they still have high nutritional requirements, and low cell biomass production yields, which makes their utilization in industrial processes difficult. Consequently, for the last three years, genetic engineering strategies were used to build recombinant xylonic acid producing strains of *E. coli, S. cerevisiae, Kluyveromyces lactis* and *Pichia kudriavzevii*[50]-[55].These microorganisms were chosen as possible hosts for presenting high growth rates, simple nutritional requirements and specially yeasts, for presenting good tolerance to inhibitors found in lignocellulosic hydrolysates, as commented above [45]. Indeed, the identification of genes from different microorganisms, that code for the enzymes involved in the conversion of xylose to xylonic acid allowed construction of strains able to produce xylonic acid with yields above 90% of theoretical maximum and at high concentrations [50]-[55].

The number of compounds naturally produced by *E. coli* is limited, and this bacterium is not a natural biofuel producer. However, advances in metabolic engineering techniques have allowed the development of strains capable of producing a big variety of biofuels from different carbon sources, such as glucose, xylose, glycerol, and fatty acids [26],[56]. An interesting example is the construction of a *E. coli* strain able to produce butanol when expressing the fermentative metabolic pathway of *Clostridia acetobutylicum*. The expression of six genes from this pathway (*thl, hbd, crt, bcd, etfAB and adhE2*) in *E. coli* was necessary to obtain a strain able to produce 139 mg L$^{-1}$butanol from glucose under anaerobic conditions [26],[56]. In an independent study, Inui and co-workers also inserted different combinations of genes from *C. acetobutylicum* butanol pathway (*thL, hbd, crt, bcd–etfB–etfA, and adhe*) in *E. coli*. The best resulting strain was able to produce 1184 mg L$^{-1}$ of butanol. Although the amount of butanol produced by the generated strains is around 10 times lower than what is obtained by *Clostridia*, these experiments show that *E. coli* is a viable host for the production of biobutanol and the power of genetic engineering [20]

# CONCLUSIONS

Several microbial-based bioprocesses are currently used in industry, and new ones should be established within the biorefinery context. To meet specific demands of the industry, which requires microbial strains able to produce fuels and chemicals from different renewable resources in high yields and productivity, researchers have been constructing and genetically improving microbial strains. The focus of these improvements can be grouped in four main categories: i) driving carbon flux towards the desired pathway, ii) increasing tolerance to toxic compounds, iii) increasing substrate uptake range, and iv) generation of new products. Thanks to the advances of genomic and molecular analysis techniques, and systemic analysis tools, microorganisms able to produce a variety of biofuels and chemicals from lignocellulose and other substrates, with production capacities in magnitudes orders higher than native ones, are currently available in the literature. Further studies concerning such microorganisms and their potential, are expected to contribute significantly to the development of bioprocesses within the biorefinery concept.

# AUTHORS' CONTRIBUTIONS

This work was carried out in collaboration between both authors. JRMA defined the review theme; BGP and JRMA wrote, read and approved the final manuscript.

# ACKNOWLEDGEMENTS

This work was financially supported by EMBRAPA (Brazilian Agricultural Research Corporation) and CNPq (Brazilian National Council for Scientific and Technological Development). BGP receive a scholarship from CAPES (Coordination for the Improvement of Higher Level Education Personnel).

# REFERENCES

1. Vijayendran B (2010) Bio products from bio refineries-trends, challenges and opportunities. Lett from Ed 7:109–115

2. [http://www.iea-bioenergy.task42-biorefineries.com/]    IEA (International Energy Agency) Bioenergy Task 42 on Biorefineries (2010)

3. Borregaard Website. [http://www.borregaard.com/] Accessed 01 Aug 2014

4. Fischer CR, Klein-Marcuschamer D, Stephanopoulos G (2008) Selection and optimization of microbial hosts for biofuels production. Metab Eng. 10:295–304

5. Van Maris AJ A, Winkler AA, Kuyper M, de Laat WT AM, van Dijken JP, Pronk JT (2007) Development of efficient xylose fermentation in Saccharomyces cerevisiae: xylose isomerase as a key component. Adv Biochem Eng Biotechnol. 108(2007):179–204

6. Goffeau A, Barrell BG, Bussey H, Davis RW, Dujon B, Feldmann H, Galibert F, Hoheisel JD, Jacq C, Johnston M, Louis EJ, Mewes HW, Murakami Y, Philippsen P, Tettelin H, Oliver SG (1996) Life with 6000 genes. Science. 274:546–567

7.  Durnin G, Clomburg J, Yeates Z, Alvarez PJJ, Zygourakis K, Campbell P, Gonzalez R (2009) Understanding and harnessing the microaerobic metabolism of glycerol in Escherichia coli. Biotechnol Bioeng. 103:148–161

8.  Mazumdar S, Clomburg JM, Gonzalez R (2010) Escherichia coli strains engineered for homofermentative production of D-lactic acid from glycerol. Appl Environ Microbiol. 76:4327–4336

9.  Causey TB, Zhou S, Shanmugam KT, Ingram LO (2003) Engineering the metabolism of Escherichia coli W3110 for the conversion of sugar to redox-neutral and oxidized products: homoacetate production. Proc Natl Acad Sci U S A. 100:825–832

10. Blankschien MD, Clomburg JM, Gonzalez R (2010) Metabolic engineering of Escherichia coli for the production of succinate from glycerol. Metab Eng. 12:409–419

11. Lee SJ, Song H, Lee SY (2006) Genome-based metabolic engineering of mannheimia succiniciproducens for succinic acid production. Appl Environ Microbiol. 72:1939–1948

12. Alsaker K, Spitzer T, Papoutsakis E (2004) Transcriptional analysis of spo0A overexpression in clostridium acetobutylicum and its effect on the cell's response to butanol stress. J Bacteriol. 186:1959–1971

13. Jiang Y, Xu C, Dong F, Yang Y, Jiang W, Yang S (2009) Disruption of the acetoacetate decarboxylase gene in solvent-producing Clostridium acetobutylicum increases the butanol ratio. Metab Eng. 11:284–291

14. Almeida JRM, Röder A, Modig T, Laadan B, Lidén G, Gorwa-Grauslund M-F (2008) NADH- vs NADPH-coupled reduction of 5-hydroxymethyl furfural (HMF) and its implications on product distribution in Saccharomyces cerevisiae. Appl Microbiol Biotechnol. 78:939–945

15. Wang X, Miller EN, Yomano LP, Zhang X, Shanmugam KT, Ingram LO (2011) Increased furfural tolerance due to overexpression of NADH-dependent oxidoreductase FucO in Escherichia coli strains engineered for the production of ethanol and lactate. Appl Environ Microbiol. 77:5132–5140

16. Yomano LP, York SW, Shanmugam KT, Ingram LO (2009) Deletion of methylglyoxal synthase gene (mgsA) increased sugar co-

metabolism in ethanol-producing Escherichia coli. Biotechnol Lett. 31:1389–1398

17. Okano K, Zhang Q, Shinkawa S, Yoshida S, Tanaka T, Fukuda H, Kondo A (2009) Efficient production of optically pure D-lactic acid from raw corn starch by using a genetically modified L-lactate dehydrogenase gene-deficient and alpha-amylase-secreting Lactobacillus plantarum strain. Appl Environ Microbiol. 75:462–467

18. Kuyper M, Harhangi H, Stave A, Winkler A, Jetten M, Delaat W, Denridder J, Opdencamp H, Vandijken J, Pronk J (2003) High-level functional expression of a fungal xylose isomerase: the key to efficient ethanolic fermentation of xylose by ? FEMS Yeast Res. 4:69–78

19. Karhumaa K, Garcia Sanchez R, Hahn-Hägerdal B, Gorwa-Grauslund M-F (2007) Comparison of the xylose reductase-xylitol dehydrogenase and the xylose isomerase pathways for xylose fermentation by recombinant Saccharomyces cerevisiae. Microb Cell Fact. 6:5

20. Inui M, Suda M, Kimura S, Yasuda K, Suzuki H, Toda H, Yamamoto S, Okino S, Suzuki N, Yukawa H (2008) Expression of clostridium acetobutylicum butanol synthetic genes in Escherichia coli. Appl Microbiol Biotechnol. 77:1305–1316

21. Steen EJ, Kang Y, Bokinsky G, Hu Z, Schirmer A, McClure A, Del Cardayre SB, Keasling JD (2010) Microbial production of fatty-acid-derived fuels and chemicals from plant biomass. Nature. 463:559–562

22. Steen EJ, Chan R, Prasad N, Myers S, Petzold CJ, Redding A, Ouellet M, Keasling JD (2008) Metabolic engineering of Saccharomyces cerevisiae for the production of n-butanol. Microb Cell Fact. 7:36

23. Tang X, Tan Y, Zhu H, Zhao K, Shen W (2009) Microbial conversion of glycerol to 1,3-propanediol by an engineered strain of Escherichia coli. Appl Environ Microbiol. 75:1628–1634

24. Blattner FR (1997) The complete genome sequence of Escherichia coli K-12. Science. 277(80):1453–1462

25. Buschke N, Schäfer R, Becker J, Wittmann C (2013) Metabolic engineering of industrial platform microorganisms for biorefinery applications–optimization of substrate spectrum and process

robustness by rational and evolutive strategies. Bioresour Technol. 135:544–554

26. Clomburg JM, Gonzalez R (2010) Biofuel production in Escherichia coli: the role of metabolic engineering and synthetic biology. Appl Microbiol Biotechnol. 86:419–434

27. Ingram LO, Conway T, Clark DP, Sewell GW, Preston JF (1987) Genetic engineering of ethanol production in Escherichia coli. Appl Environ Microbiol. 53:2420–2425

28. Chemicals Technology. [http:/ / www.chemicals-technology. com/ projects/ stan-mayfield-biorefinery-pilot-pla nt/ ] The Stan Mayfield Pilot Plant Website. Accessed 01 Aug 2014

29. Alper H, Stephanopoulos G (2009) Engineering for biofuels: exploiting innate microbial capacity or importing biosynthetic potential? Nat Rev Microbiol. 7:715–723

30. Börgel D, van den Berg M, Hüller T, Andrea H, Liebisch G, Boles E, Schorsch C, van der Pol R, Arink A, Boogers I, van der Hoeven R, Korevaar K, Farwick M, Köhler T, Schaffer S (2012) Metabolic engineering of the non-conventional yeast Pichia ciferrii for production of rare sphingoid bases. Metab Eng. 14:412–426

31. Brown NA, de Castro P, de Castro BPF, Savoldi M, Buckeridge MS, Lopes ML, de Lima Paullilo SC, Borges EP, Amorim HV, Goldman MHS, Bonatto D, Malavazi I, Goldman GH (2013) Transcriptional profiling of Brazilian Saccharomyces cerevisiae strains selected for semi-continuous fermentation of sugarcane must. FEMS Yeast Res. 13:277–290

32. Mojzita D, Wiebe M, Hilditch S, Boer H, Penttilä M, Richard P (2010) Metabolic engineering of fungal strains for conversion of D-galacturonate to meso-galactarate. Appl Environ Microbiol. 76:169–175

33. Anderlund M, Nissen TL, Nielsen J, Villadsen J, Rydström J, Hahn-Hägerdal B, Kielland-Brandt MC (1999) Expression of the Escherichia coli pntA and pntB genes, encoding nicotinamide nucleotide transhydrogenase, in Saccharomyces cerevisiae and its effect on product formation during anaerobic glucose fermentation. Appl Environ Microbiol. 65:2333–2340

34. Nissen TL, Kielland-Brandt MC, Nielsen J, Villadsen J (2000) Optimization of ethanol production in Saccharomyces cerevisiae

by metabolic engineering of the ammonium assimilation. Metab Eng. 2:69–77

35. Roca C, Nielsen J, Olsson L (2003) Metabolic engineering of ammonium assimilation in xylose-fermenting Saccharomyces cerevisiae improves ethanol production. Appl Environ Microbiol. 69:4732–4736

36. Almeida JRM, Bertilsson M, Hahn-Hägerdal B, Lidén G, Gorwa-Grauslund M-F (2009) Carbon fluxes of xylose-consuming Saccharomyces cerevisiae strains are affected differently by NADH and NADPH usage in HMF reduction. Appl Microbiol Biotechnol. 84:751–761

37. Alper H, Stephanopoulos G (2007) Global transcription machinery engineering: a new approach for improving cellular phenotype. Metab Eng. 9:258–267

38. Hahn-Hägerdal B, Karhumaa K, Fonseca C, Spencer-Martins I, Gorwa-Grauslund MF (2007) Towards industrial pentose-fermenting yeast strains. Appl Microbiol Biotechnol. 74:937–953

39. Almeida JR, Modig T, Petersson A, Hähn-Hägerdal B, Lidén G, Gorwa-Grauslund MF (2007) Increased tolerance and conversion of inhibitors in lignocellulosic hydrolysates bySaccharomyces cerevisiae. J Chem Technol Biotechnol. 82:340–349

40. Almeida JRM, Runquist D, Sànchez i Nogué V, Lidén G, Gorwa-Grauslund MF (2011) Stress-related challenges in pentose fermentation to ethanol by the yeast Saccharomyces cerevisiae. Biotechnol J. 6:286–299

41. Heer D, Sauer U (2008) Identification of furfural as a key toxin in lignocellulosic hydrolysates and evolution of a tolerant yeast strain. Microb Biotechnol. 1:497–506

42. Hughes SR, Gibbons WR, Bang SS, Pinkelman R, Bischoff KM, Slininger PJ, Qureshi N, Kurtzman CP, Liu S, Saha BC, Jackson JS, Cotta M, Rich JO, Javers JE (2012) Random UV-C mutagenesis of Scheffersomyces (formerly Pichia) stipitis NRRL Y-7124 to improve anaerobic growth on lignocellulosic sugars. J Ind Microbiol Biotechnol. 39:163–173

43. Liu ZL, Slininger PJ, Dien BS, Berhow MA, Kurtzman CP, Gorsich SW (2004) Adaptive response of yeasts to furfural and 5-hydroxymethylfurfural and new chemical evidence for HMF

conversion to 2,5-bis-hydroxymethylfuran. J Ind Microbiol Biotechnol. 31:345–352

44.  Hou X, Yao S (2012) Improved inhibitor tolerance in xylose-fermenting yeast Spathaspora passalidarum by mutagenesis and protoplast fusion. Appl Microbiol Biotechnol. 93:2591–2601

45.  Almeida JRM, Bertilsson M, Gorwa-Grauslund MF, Gorsich S, Lidén G (2009) Metabolic effects of furaldehydes and impacts on biotechnological processes. Appl Microbiol Biotechnol. 82:625–638

46.  Almeida JRM, Fávaro LCL, Quirino BF (2012) Biodiesel biorefinery: opportunities and challenges for microbial production of fuels and chemicals from glycerol waste. Biotechnol Biofuels. 5:48

47.  Ferreira-Leitão V, Perrone CC, Rodrigues J, Franke APM, Macrelli S, Zacchi G (2010) An approach to the utilisation of CO2 as impregnating agent in steam pretreatment of sugar cane bagasse and leaves for ethanol production. Biotechnol Biofuels. 3:7

48.  Hahn-Hägerdal B, Karhumaa K, Jeppsson M, Gorwa-Grauslund MF (2007) Metabolic engineering for pentose utilization in Saccharomyces cerevisiae. Adv Biochem Eng Biotechnol. 108:147–177

49.  Werpy T, Petersen G, Aden A, Bozell J (2004) Top value added chemicals from biomass. Volume 1-Results of screening for potential candidates from sugars and synthesis gas

50.  Toivari MH, Ruohonen L, Richard P, Penttilä M, Wiebe MG (2010) Saccharomyces cerevisiae engineered to produce D-xylonate. Appl Microbiol Biotechnol. 88:751–760

51.  Nygård Y, Toivari MH, Penttilä M, Ruohonen L, Wiebe MG (2011) Bioconversion of d-xylose to d-xylonate with Kluyveromyces lactis. Metab Eng. 13:383–391

52.  Liu H, Valdehuesa KNG, Nisola GM, Ramos KRM, Chung W-J (2012) High yield production of D-xylonic acid from D-xylose using engineered Escherichia coli. Bioresour Technol. 115:244–248

53.  Toivari M, Vehkomäki M-L, Nygård Y, Penttilä M, Ruohonen L, Wiebe MG (2013) Low pH D-xylonate production with Pichia kudriavzevii. Bioresour Technol. 133:555–562

54. Cao Y, Xian M, Zou H, Zhang H (2013) Metabolic engineering of Escherichia coli for the production of xylonate. PLoS One. 8:e67305

55. Toivari MH, Nygård Y, Penttilä M, Ruohonen L, Wiebe MG (2012) Microbial D-xylonate production. Appl Microbiol Biotechnol. 96:1–8

56. Atsumi S, Hanai T, Liao JC (2008) Non-fermentative pathways for synthesis of branched-chain higher alcohols as biofuels. Nature. 451:86–89

# A Potential Source for Cellulolytic Enzyme Discovery and Environmental Aspects Revealed Through Metagenomics of BrazilianMangroves

Claudia Elizabeth Thompson[1,2], Walter Orlando Beys-da-Silva[2],
Lucélia Santi[2], Markus Berger[2], Marilene Henning Vainstein[2],
Jorge Almeida Guima rães[2], and Ana Tereza Ribeiro
Vasconcelos[1]

[1]Laboratório Nacional de Computação Científica, Rio de Janeiro 25651070, Brazil

[2]Centro de Biotecnologia, Universidade Federal do Rio Grande do Sul, Rio Grande do Sul 91501070, Brazil

# ABSTRACT

The mangroves are among the most productive and biologically important environments. The possible presence of cellulolytic enzymes and microorganisms useful for biomass degradation as well as taxonomic and functional aspects of two Brazilian mangroves were evaluated using cultivation and metagenomic approaches. From a total of 296 microorganisms with visual differences in colony morphology and growth (including bacteria, yeast and filamentous fungus), 179 (60.5%) and 117 (39.5%) were isolated from the Rio de Janeiro (RJ) and Bahia (BA) samples, respectively. RJ metagenome showed the higher number of microbial isolates, which is consistent with its most conserved state and higher diversity. The metagenomic sequencing data showed similar predominant bacterial phyla in the BA and RJ mangroves with an abundance of Proteobacteria (57.8% and 44.6%), Firmicutes (11% and 12.3%) and Actinobacteria (8.4% and 7.5%). A higher number of enzymes involved in the degradation of polycyclic aromatic compounds were found in the BA mangrove. Specific sequences involved in the cellulolytic degradation, belonging to cellulases, hemicellulases, carbohydrate binding domains, dockerins and cohesins were identified, and it was possible to isolate cultivable fungi and bacteria related to biomass decomposition and with potential applications for the production of biofuels. These results showed that the mangroves possess all fundamental molecular tools required for building the cellulosome, which is required for the efficient degradation of cellulose material and sugar release.

# INTRODUCTION

Mangroves are coastal ecosystems that are found in tropical and subtropical regions and that cover between 60 and 75% of the world's coasts (Holguin et al. 2001). These ecosystems have unique characteristics, including brackish water, particular sediments (muddy soil), and a specific population of animals and plants. They represent a complex and dynamic ecosystem that varies in terms of its salinity, water levels and nutrient availability during the seasons (Gomes et al. 2008; Gonzalez-Acosta et al. 2006). Mangroves are among the most productive and biologically important environments (Giri et al.

2011). They play a critical role in water filtering and establishing and expanding the coastal line, and they are of fundamental importance in maintaining the food chain and carbon cycle for human, coastal and marine communities (Giri et al. 2011; Gomes et al. 2008).

Mangrove forests are often located in urban areas touched by constant anthropogenic activity, which is thought to be the major cause of coastal wetland deterioration. Brazil is the third most mangrove-rich country, with about 7% of the world's total area of mangroves (Giri et al.2011). However, these areas are under constant flux due to both natural and anthropogenic forces. The mangrove forests are under immense pressure from clear-cutting, land-use changes, hydrological alterations, chemical spills and climate changes. Due to their singular characteristics and nutrient availability, mangroves have a remarkable influence on microbial communities, and their activities are responsible for several processes in organic matter degradation and most of the carbon flow in their sediments (Holguin et al. 2001).

Importantly, mangrove forests are believed to play a role in the uptake and preservation of polycyclic aromatic hydrocarbons (PAHs). PAHs are contaminants with both known and suspected carcinogenic, toxic, and mutagenic properties. They enter the aquatic environments through atmospheric deposition, municipal effluents, industrial wastewater, and oil spillage (Zhou et al. 2008). The mangroves are exposed to anthropogenic PAH contamination from tidal water, river water and land-based sources. Consequently, they supply organic matter to the coastal waters and influence global biogeochemical nutrient cycling (Sun et al. 2012).

The adaptation of bacterial species to the mangrove ecosystem indicates a potential source of biotechnological resources. The mangroves could provide a rich resource for the discovery of new bacterial and fungal species that produce enzymes and molecules that could be used for human life, agriculture, industry and bioremediation (Dourado et al. 2012; Dias et al. 2009). The ability to convert lignocellulosic substrates in nutrients of low complexity is crucial for the carbon cycle and microbial survivability, especially in environments with large amounts of these substrates (Medie et al. 2012). Lignocellulosic materials are very difficult to degrade due to their dense and compact structural features, which are related to the protection of plants against microbial attack (Kumar et al. 2008). Microorganisms that specialize in vegetable biomass degradation might be found in specific environments, such as the mangrove ecosystem, where such biomass is abundant. These microorganisms, which have an enormous ecological relevance, may also have a biotechnological use in the production of biofuels.

Unlike other strategies used for novel enzyme and microorganism identification, metagenomic analysis has clear benefits as a powerful alternative to culture-dependent methods. High-throughput pyrosequencing is a new tool for the study of microbial ecology and can reveal the taxonomic diversity of specific environments at a high resolution (Zhu et al. 2011). It is becoming the most powerful tool for studying uncultured microbes and novel enzymes with biotechnological potential (Duan et al. 2010; Brulc et al. 2009). A metagenomic pyrosequencing-based approach for the isolation of cultivable microorganisms could accelerate the assessment of new enzymes and molecules and their possible application in biotechnological purposes. This approach may also be useful in the study of biodiversity and provide evidence about environmental interference.

The objective of this work was to evaluate the possible presence of cellulolytic enzymes and microorganisms useful for biomass degradation as well as taxonomic and functional aspects of two Brazilian mangroves. Using a metagenomic approach, the specific sequences and microorganisms involved in cellulolytic degradation were identified, and several cellulolytic microorganisms were isolated from the mangrove samples. The data provided here demonstrate the biotechnological potential of this unique but endangered environment.

# MATERIAL AND METHODS

### Sample Collection and DNA Extraction

Samples were collected from two different Brazilian mangroves: sample RJ was collected from a preserved mangrove in the Rio de Janeiro state Ilha Grande during October of 2010 (23° 10′ 11.852″ S, 44° 17′ 0.2″ W), and sample BA was collected from a periurban mangrove area with anthropogenic action in the vicinity of the Bahia state Porto Seguro during November of 2010 (16° 35′ 17.225″ S, 39° 5′ 30.642″ W). In relation to the BA sample, no specific permissions were required according to the Brazilian government rules (resolution n. 21, August 31, 2006 - Ministry of Environment). The RJ sample was collected in accordance with the Brazilian law (IN 154/2007 IBAMA, Brazilian Institute of Environment and Renewable Natural Resources) and we confirm that the field studies did not involve endangered or protected species.

The samples (50 g) were collected from the soil surface (0–10 cm) using sterilized spatulas and sterilized 50 mL plastic tubes. The samples were maintained on ice during transportation until processing. Some physical-chemical and environmental characteristics of the samples were determined, such as their pH, temperature, and salinity. The samples collected were manually shaken for 2 minutes in 100 mL sterile Erlenmeyer flasks and 10 g were taken for DNA extraction. DNA was extracted from 10 g of each soil replicate using the PowerMax™ Soil DNA Isolation Kit (MoBio Laboratories, Inc., Carlsbad, CA, USA) according to the manufacturer's recommendations. The quantity and quality of the resulting DNA were verified with a NanoDrop (Thermo Scientific, Wilmington, DE, USA) spectrophotometer at an absorbance ratio of 260/280 nm and 260/230 nm and were confirmed by electrophoresis on a 1% agarose gel. The DNA samples were then diluted to a concentration of 50 ng/µL.

## The Isolation of Cellulolytic Microorganisms

Approximately 5 g of soil collected from both mangroves were inoculated in 200 mL of minimal medium MM (peptone 0.3%, $K_2HPO_4$ 0.05%, $MgSO_4$ 0.05%) containing sugar cane bagasse 3% as the carbon source. The culture flasks were shaken for 5 min and maintained without agitation during 3 days at 28°C; the flasks were then incubated for 10 days at 28°C and 100 rpm. Samples of 50 µL each were collected in two-day intervals and cultured on Petri dish plates containing four different types of media: minimal medium with sugar cane bagasse 1% (M1), minimal medium with carboxymethyl cellulose (CMC) 1% (M2), Luria-Bertani (LB) medium (M3), and Sabouraud (M4) medium. Microorganisms with different colony morphologies were isolated and used to generate pure cultures, which were subsequently maintained at 4°C.

The cellulolytic ability of each isolate was tested by inoculation into 3 mL tubes of minimal medium with CMC plus sugarcane bagasse, pinus (*Pinus elliottii*) powder, or medium-density fiberboard (MDF) powder at 1%, as the main carbon source. After incubation for 5 days at 28°C and 150 rpm, the isolates that grew in at least in two of these substrates were also tested in a CMC-Congo red plate assay (Teather et al. 1982).

# DNA Pyrosequencing and Sequence Processing

The preparation of two libraries was accomplished according to instructions from the Rapid Library Preparation Method Manual - GS FLX Titanium Series (454-Roche). The first library was made from 500 ng of DNA extracted from the RJ mangrove soil sample and another from 500 ng of DNA extracted from the BA mangrove soil sample. The titration, emulsion PCR, and sequencing steps were performed according to the manufacturer's instructions. A two-region 454 sequencing run was performed on a 70x75 PicoTiterPlate (PTP) using the Genome Sequencer FLX System (Roche). Each region was loaded with one of the library preparations.

The artificially replicated sequences that were an artifact of the 454-based pyrosequencing technique were identified and eliminated using the Replicates software (Gomes-Alvarez et al.2009). The remaining reads were further filtered with the LUCY program to remove short sequences (less than 180 bp) and sequences with a phred quality $\leq 20$ (Chou and Holmes2001).

The Newbler Assembler software version 2.5.3 was used to perform the assembly procedures, with a "-rip" flag that allows for the output of each read in a contig. Reads identified by the GS De Novo Assembler as problematic (Partial, Repeat, Outlier, TooShort) or with High-Quality Discrepancies (HQD) were filtered from the dataset. Subsequently, a new cycle of assembling and filtering was performed. These steps were repeated until the problematic and high-quality discrepancy reads were decreased below the cutoff threshold of 1% of the total number of reads filtered out at the first assembly step.

A total of 621,748 and 647,534 reads were produced by the 454-Titanium pyrosequencer after raw data processing for the BA and the RJ mangroves, respectively. The Replicates tool identified and removed 58,121 (BA) and 92,644 (RJ) replicate reads. From the remaining 563,627 and 554,890 reads, the LUCY tool removed another 60,608 and 60,689 low-quality reads, such that 503,019 (BA) and 494,201 (RJ) sequences were deemed appropriate for further taxonomic and functional categorization (Table 1). As a result of the assembly, 419,591 and 481,593 reads remained as singletons, while 72 and 29 contigs ≥ 500 bp were formed for BA and RJ, respectively. The average contig size was 23,788 bp for the RJ and 1,002 bp for the BA metagenomes.

**Table 1:** Summary of metagenomic data obtained from the mangrove micro-biomes

| Parameters | Rio | Bahia |
|---|---|---|
| MG-RAST ID | 44852183 | 44852193 |
| No. of sequences | 494,201 | 503,019 |
| Avg. length (bp) | 327 ± 110* | 328 ± 111* |
| Total length (bp) | 161,854,245 | 165,387,021 |
| Predicted proteins† | 488,630 | 450,087 |
| Assigned reads | 242,868 | 261,482 |
| LCA¥ | | |
| *Bacteria* (%) | 206,733 (94,3) | 273,078 (97,36) |
| *Archaea* (%) | 10,041 (4,6) | 4,623 (1,65) |
| *Eukarya* (%) | 1,559 (0,7) | 1,018 (0,36) |
| *Viruses* (%) | 297 (0,1) | 29 (0,01) |
| Unclassified (%) | 649 (0,3) | 1,740 (0,62) |
| MEGAN | | |
| No. of sequences | 494,201 | 503,019 |
| Assigned reads | 293,374 | 340,668 |
| *Bacteria* (%) | 252,395 (93,6) | 313,633 (97,3) |
| *Archaea* (%) | 13,173 (4,88) | 5,686 (1,76) |
| *Eukarya* (%) | 3,295 (1,22) | 2,805 (0,87) |
| *Viruses* (%) | 554 (0,2) | 57 (0,01) |
| Unclassified (%) | 269 (0,1) | 182 (0,06) |

*After duplicate removal, splitting and trimming of sequence reads.

†Predicted protein coding regions assigned an annotation using at least one of protein databases (M5NR) in MG-RAST server.

¥Lowest common ancestor (LCA) using 1e-5 cutoff, 60% minimum identity, and a minimum alignment lenght cutoff of 15.

Thompson *et al.*

Thompson *et al. AMB Express* 2013 **3**:65, doi:10.1186/2191-0855-3-65

Metagenome data reported in this paper have been deposited into the GenBank database (PRJNA186597) and on the MG-RAST server (4485218.3, 4485219.3, 4485981.3, and 4485982.3).

# Taxonomic Distribution Analysis

The taxonomic distribution of each read was performed with the MEGAN4 (Huson et al.2011) software, which compares the given reads against a database of reference sequences by performing a search using the BLASTX algorithm (Altschul et al. 1997) against the NCBI-NR protein database. The MG-RAST server (Meyer et al. 2008) was also used for the taxonomic analysis. The 16S rRNA identification was accomplished using the Meta_RNA software (Huang et al. 2009). The sequences were subsequently classified with RDP Classifier software (Wang et al. 2007) based on the RDP Database (Cole et al. 2009).

# Functional Metagenomic Analysis

The MEGAN4 software was used to perform a functional analysis using the SEED (Overbeek et al. 2005) and KEGG (Kanehisa et al. 2012) databases. Each read was related to its SEED functional role using the best BLAST score to protein sequences without known functional roles. A similar procedure was used to match each read to a KEGG orthology (KO) accession number. To study specific metabolic pathways, the BA and RJ metagenomes were compared using 1e-05 as the maximum e-value cutoff, with a minimum identity of 60% and a minimum alignment length of 15. A functional comparison against the public metagenomes was also performed on the MG-RAST server.

Carbohydrate-active enzyme domains assigned by the CAZy database (Cantarel et al. 2009) were searched for in the predicted protein sequences (Rho et al. 2010; Zhu et al. 2010) using HMMER 3.0 software (Finn et al. 2011) and the Pfam 26.0 database (Punta el al. 2012), with a cutoff of $E \leq 1e-4$.

All contig sequences were analyzed and functionally annotated using the System for Automated Bacterial Integrated Annotation (SABIA) (Almeida et al. 2004). According to the automatic annotation criteria, a given ORF was considered "valid" if it had BlastP hits on the KEGG, NCBI-nr or UniProtKB/Swiss-Prot databases, if it had subject and query coverage of $\geq 60\%$, and if it had positives of $\geq 60\%$. ORFs that had no BlastP hits on the NCBI-nr, KEGG, UniProtKB/Swiss-Prot, TCDB or Interpro databases or that were excluded by the above criteria were considered "hypothetical".

# Comparative Metagenomic Analysis

The mangrove samples were compared with other metagenomes using the SEED subsystem content as a comparative metric (Suen et al. 2010 Tringe et al. 2005). Metagenomes from host-associated, water, and soil samples were evaluated using the MG-RAST server. The parameters for inclusion were a maximum e-value cutoff of 1e-05, a minimum identity of 60%, and a minimum alignment length of 15.

Trends in the abundance of the SEED subsystem were examined using Principal Component Analysis (PCA) and hierarchical clustering (Willner et al. 2009). The PCA method is a reduction/ordination technique that clusters the samples based on variations extracted from their normalized abundance profiles (Meyer et al. 2008). This analysis was performed on normalized data, using the bray-curtis dissimilarity.

# Statistical Analyses

The MG-RAST taxonomic and functional profiles were analyzed using the Statistical Analyses of Metagenomic Profiles (STAMP) software (Parks and Beiko 2010) to detect biologically relevant differences in the relative proportions of the classified sequences. The two-sided Fisher's exact test with Storey's FDR method for multiple test correction was employed to analyze the data sets. All unclassified reads were removed from analyses. The most important taxa were filtered according to their q-values (0.01) and only those categories with more than a 2-fold ratio between the proportions or with difference between the proportions of at least 5% were used.

Statistical tests on the taxonomic data were also performed with MEGAN. The BA and RJ counts were normalized to produce data sets of 100,000 reads. Subsequently, MEGAN was used to apply a directed homogeneity test in order to highlight the significant differences in the mangrove comparisons. The highlighting thickness is logarithmically proportional to its significance; that is, the thickness is an integer value of $2\log x$ when $P = 1.0e^{x}$ (Mitra et al.2009). Multiple testing correction analysis was not applied, and all unassigned reads were ignored.

# RESULTS

Sample Description

Two different mangrove locations with distinct features were chosen: one in the Southeast (RJ) and the other in the Northeast (BA) of the country. The RJ mangrove is located in the city of Angra dos Reis, in Rio de Janeiro state, on a natural reserve island called Ilha Grande. Ilha Grande was created by the Brazilian government on August 25, 1978, and is considered to be highly preserved. It is 3,600 ha in size and contains both mangroves and other coastal ecosystems, including the Atlantic rainforest of subtropical southeast Brazil. The BA mangrove is located in the city of Porto Seguro, in the Trancoso district of the Bahia state, in a tropical region of the continent. It also contains both coastal and restinga ecosystems. Despite its allocation in a hard-to-access site, the area is close to several small rural properties and a few luxury beach hotels and resorts. Consequently, this area is now increasingly threatened by urban development. Therefore, the BA mangrove is considered to be at a medium grade of degradation (Gomes et al. 2008). Although these two mangrove systems differ in terms of their location (island and continent), anthropic action and preservation state, they are similar in terms of their physical-chemical parameters. They both have a humid climate with average temperatures of 24°C and average salinity levels of 1.5%. The RJ mangrove has a neutral pH of 7.0, while the BA mangrove has a slightly acidic pH of 6.5. The samples were collected from the soil surface and the intertidal zone to access different soil layers along with their associated microorganisms.

## The Isolation and Selection of Cultivable Microorganisms

Samples from the RJ and BA mangroves were used to isolate cultivable microorganisms in four different media. From a total of 296 microorganisms with visual differences in colony morphology and growth (including bacteria, yeast and filamentous fungus), 179 (60.5%) and 117 (39.5%) were isolated from the RJ and BA samples, respectively (Figure 1A).

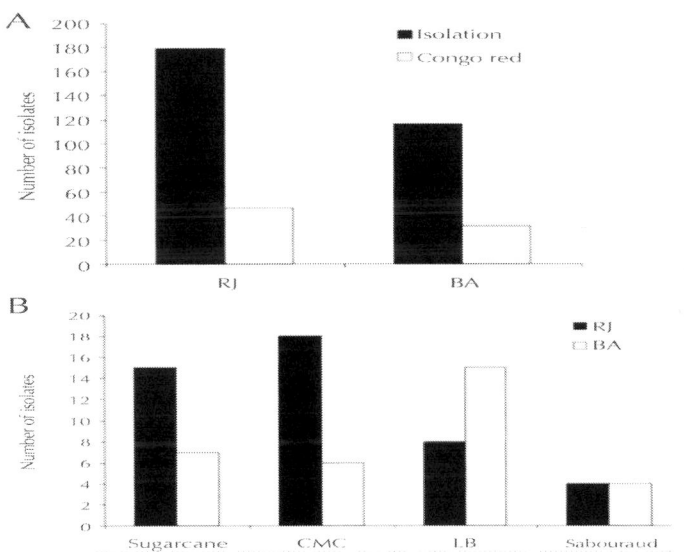

**Figure 1:** Cellulolytic microorganisms isolated from Brazilian mangroves. (A) Total number of microbial isolates and positives for CMC-Congo red plate assay. (B) Number of microbial isolates corresponding to each culture medium in the first step of isolation.

To determine the ability of these isolates to degrade cellulosic compounds, all microorganisms were grown in a CMC-congo red assay. Here, 77 microorganisms were found to be positive, and 58.4%

of these were isolated from the RJ mangrove. In addition, 73% of the CMC-congo red positive microorganisms were initially isolated from mangrove samples through media containing cellulosic compounds (sugarcane bagasse or CMC) (Figure 1B). Twenty-one filamentous fungi with cellulolytic potential were isolated.

# The Biodiversity and General Characteristics of the Mangrove Metagenome

A large-scale metagenomic analysis was used to profile the microbiota of two Brazilian mangroves and to identify their functional attributes. A total of 503,019 and 494,201 reads from BA and RJ, respectively, were submitted for taxonomic and functional analysis. After applying quality control measures and removing any artificial duplicate reads (ADRs) on the MG-RAST server, the 491,224 (98.7%) reads from the BA library produced a total of 450,087 predicted protein-coding regions. Similarly, a total of 488,630 protein-coding regions were predicted from the RJ mangrove metagenome (Table 1). Of these sequences, 772 (BA) and 457 (RJ) were identified as 16S rDNAs using the MG-RAST server against RDP. This result corresponds to 0.15% and 0.09% of the total sequences obtained from the BA and RJ samples, respectively.

For the BA metagenome, 261,482 (58.1%) of the 450,087 predicted proteins were assigned an annotation, and 188,605 (41.9%) had no significant similarity to any protein in the databases and were therefore considered orphan sequences. In total, 243,812 (93.2%) sequences were functionally categorized. In the RJ sample, 242,868 (49.7%) sequences were assigned an annotation, and 245,762 (50.3%) were found to have no significant similarities to anything in the protein databases. Of the annotated sequences, 224,512 (92.4%) were assigned to functional categories. The average read length for the BA sample was $328 \pm 111$ bp, with a mean GC content of $54 \pm 10\%$, and the average read length for the RJ sample was $327 \pm 110$ bp with a mean GC content of $55 \pm 10\%$ (Table 1).

The MEGAN results assigned annotations to 340,668 reads from the BA mangrove, which corresponds to 67.7% of the total reads, 162,169 (32.2%) of which are orphan sequences. For the RJ sample, 293,374 (59.3%) reads had significant similarities to hits within the protein databases, whereas 200,458 (40.5%) were unassigned reads. The algorithm implemented by MEGAN assigned more sequences when compared to MG-RAST, and it was able to identify a higher number of sequences related to *Bacteria*, *Archaea*, *Eukarya*, and *Viruses* (Table 1).

The alpha diversity was calculated for both samples, and the results indicated that the RJ mangrove (alpha-diversity = 974.385 species) is more diverse than the BA mangrove (alpha-diversity = 847.721 species). The index of alpha-diversity summarizes the diversity of organisms in a sample and is estimated according to the distribution of species-level annotations.

# The Analysis of Mangrove-Associated Microbiota

At the domain level, *Bacteria* were more abundant than *Archaea* in the two mangrove samples. The MG-RAST analysis indicated that the percentage of sequences affiliated with each taxa were similar in the BA and RJ mangroves, with an abundance of *Proteobacteria* (57.8% and 44.6%), *Firmicutes* (11% and 12.3%),*Actinobacteria* (8.4% and 7.5%), *Bacteroidetes* (4.1% and 4.3%), *Chloroflexi* (3.2% and 6.4%), *Cyanobacteria* (2.2% and 5.8%), *Euryarchaeota* (2.2% and 4.5%), and *Planctomycetes*(1.6% and 1.9%). The statistical test applied by MEGAN revealed that the proportional difference in reads assigned to *Chloroflexi*, *Cyanobacteria*, *Planctomycetes*, *Proteobacteria*, and *Chlamydiae/Verrucomicrobia* was highly significant between the BA and RJ mangroves, with the RJ sample having a higher number of reads assigned to those taxa. Moreover, the*Actinobacteria*, *Fibrobacteres/ Acidobacteria*, and *Firmicutes* groups were significantly more highly represented in the BA mangrove (Figure 2A).

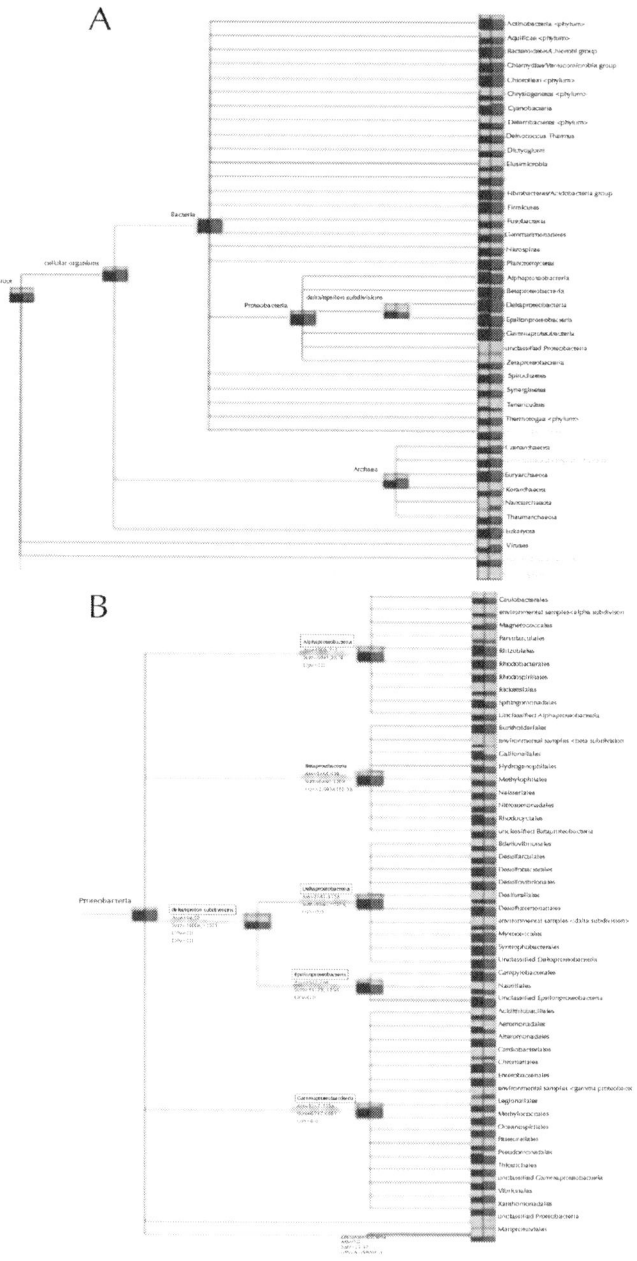

**Figure 2:** Analysis of the composition of mangrove microbial communities, showing the taxonomic diversity of metagenomic sequences, computed by MEGAN based on a BLASTX using an e-value cutoff of 1e-5 comparison for

Bahia (blue) e Rio de Janeiro (red). (A) *Bacteria, Archaea, Eukaryota,* and *Viruses*. (B)*Proteobacteria*. The assigned and summarized reads for*Alphaproteobacteria, Betaproteobacteria, Deltaproteobacteria,Epsilonproteobacteria,* and *Zetaproteobacteria* nodes are displayed. Yellow highlighting on the left side of a node indicates that the up-test of directed homogeneity test showed a significant difference. The thickness of the highlighting is logarithmically proportional to the significance. The size of the bars is scaled logarithmically to represent the number of reads assigned to each taxon. The UPv labels indicate the P-values associated with the up parts of the directed homogeneity test.

In terms of the phylogenetic classification within *Proteobacteria* obtained via MG-RAST, the most frequent class detected was *Gammaproteobacteria* (13.5% in BA and 13.6% in RJ), followed by *Epsilonproteobacteria* (11.6%), *Alphaproteobacteria* (11.5%),*Deltaproteobacteria* (10.9%), and *Betaproteobacteria* (9.9%) in the BA sample, and followed by *Deltaproteobacteria* (12.5%), *Betaproteobacteria* (8.2%), *Alphaproteobacteria* (7.9%), and*Epsilonproteobacteria* (2.1%) in the RJ sample. The statistical analyses from MEGAN indicated that the BA mangrove had significantly more reads in *Alphaproteobacteria,Betaproteobacteria,* and *Epsilonproteobacteria,* whereas the RJ mangrove had significantly more reads in *Gammaproteobacteria, Deltaproteobacteria,* and *Zetaproteobacteria*(Figure 2B).

At the genus level, *Pseudomonas* (4.3%), *Campylobacter* (2.9%), and *Geobacter* (2.2%) were found to be the main bacterial groups in the BA sample, as measured according to the MG-RAST results. In contrast, *Geobacter* (2.2%), *Roseiflexus* (1.9%), and *Burkholderia* (1.8%) were found to be the main bacterial groups in the RJ sample.

The BA and RJ mangroves were statistically different in terms of their *Archaea* abundance, with the RJ sample having a higher abundance (Figure 2A). The majority of Archaeal EGTs corresponded to methanogenic classes, with the largest proportion corresponding to the*Euryarchaeota* (Figure 2A). The genomes of the most representative strains from each mangrove are quite different (Table 2). Anaerobic bacteria are more heavily represented in the RJ mangrove, several of which belong to the *Deltaproteobacteria* genus. In contrast, the BA mangrove has several reads from *Actinobacteria,* which are not heavily represented in the RJ mangrove samples. Moreover, some Euryarchaeota genomes were identified only in the RJ sample (Table 2).

**Table 2:** Sequenced genomes with most hits to the mangrove metagenomes

RJ Mangrove

| Genome | Kingdom/Group | Gram Strain | Oxygen requeriment | Motility | Pathogenic | Habitat | Salinity | Number of reads |
|---|---|---|---|---|---|---|---|---|
| Syntrophobacter fumaroxidans MPOB | Bacterial Deltaproteobacteria | Negative | Anaerobic | No | No | Aquatic | | 1,339 |
| Desulfatibacillum alkenivorans AK-01 | Bacterial Deltaproteobacteria | Negative | Anaerobic | No | No | Aquatic | | 1,007 |
| Syntrophus aciditrophicus SB | Bacterial Deltaproteobacteria | Negative | Anaerobic | No | | Multiple | | 584 |
| Thioalkalivibrio sulfidophilus HL-EbGr7 | Bacterial Gammaproteobacteria | Negative | Aerobic | Yes | No | Specialized | Moderate halophilic | 472 |
| Roseiflexus sp. RS-1 | Bacteria/Chloroflexi | Negative | Facultative | Yes | No | Specialized | | 421 |
| Gemmatimonas aurantiaca T-27 | Bacteria | Negative | Aerobic | Yes | No | | | 374 |
| Opitutus terrae PB90-1 | Bacteria | | Anaerobic | Yes | No | Aquatic | | 290 |
| Thermomicrobium roseum DSM 5159 | Bacteria/Chloroflexi | Negative | Aerobic | No | No | Specialized | | 224 |
| Geobacter uraniireducens Rf4 | Bacterial Deltaproteobacteria | Negative | Microaerophilic | Yes | No | Multiple | | 222 |
| Thiobacillus denitrificans ATCC 25259 | Bacterial Betaproteobacteria | Negative | Facultative | Yes | No | Multiple | | 214 |
| Methanosarcina acetivorans C2A | Archaea/Euryarchaeota | | Anaerobic | No | No | Aquatic | | 179 |
| Methanoculleus marisnigri JR1 | Archaea/Euryarchaeota | Negative | Anaerobic | Yes | No | Aquatic | | 170 |

BA Mangrove

| Genome | Kingdom/Group | Gram Strain | Oxygen requeriment | Motility | Pathogenic | Habitat | Salinity | Number of reads |
|---|---|---|---|---|---|---|---|---|
| *Thiobacillus denitrificans* ATCC 25259 | *Bacteria/ Betaproteobacteria* | Negative | Facultative | Yes | No | Multiple | | 984 |
| *Syntrophobacter fumaroxidans* MPOB | *Bacteria/ Deltaproteobacteria* | Negative | Anaerobic | No | No | Aquatic | | 911 |
| *Desulfatibacillum alkenivorans* AK-01 | *Bacteria/ Deltaproteobacteria* | Negative | Anaerobic | No | No | Aquatic | | 771 |
| *Syntrophus aciditrophicus* SB | *Bacteria/ Deltaproteobacteria* | Negative | Anaerobic | No | | Multiple | | 492 |
| *Gemmatimonas aurantiaca*T-27 | *Bacteria* | Negative | Aerobic | Yes | No | | | 326 |
| *Opitutus terrae* PB90-1 | *Bacteria* | Negative | Anaerobic | Yes | No | Aquatic | | 250 |
| *Geobacter uraniireducens* Rf4 | *Bacteria/ Deltaproteobacteria* | Negative | Microaerophilic | | | Multiple | | 230 |
| *Roseiflexus* sp. RS-1 | *Bacteria/Chloroflexi* | Negative | Facultative | Yes | No | Specialized | | 220 |
| *Rubrobacter xylanophilus* DSM 9941 | *Bacteria/Actinobacteria* | Positive | Aerobic | No | No | Specialized | | 211 |
| *Nocardioides* sp. JS614 | *Bacteria/Actinobacteria* | Positive | Aerobic | No | No | Terrestrial | | 157 |
| *Thioalkalivibrio sulfidophilus* HL-EbGr7 | *Bacteria/ Gammaproteobacteria* | Negative | Aerobic | Yes | No | Specialized | | 151 |

*Thompson et al.*

Thompson et al. AMB Express 2013 **3**:65, doi:10.1186/2191-0855-3-65

According to the MEGAN results, the majority of the eukaryotic EGTs (40.1 and 37.2% for BA and RJ, respectively) were more similar to the *Opisthokonta*, *Stramenopiles* (21.8 and 32.2%), and *Viridiplantae* (24.3 and 18.7%, respectively). However, there was no significant difference between the mangroves in terms of either their eukaryotic or their viral sequences.

# Functional Metagenomic Analyses

Subsystem-based annotations (SEED) were performed to analyze the metabolic and physiological conditions in the metagenome of the mangrove samples (Figure 3). A wide variety of environmental gene tags (EGTs) from several metabolic routes were found. The mangroves were found to have a similar profile in terms of the percentages of genes from all of the SEED terms identified. Moreover, the MG-RAST (Figure 3) and MEGAN analyses gave similar patterns for both groups of mangrove samples.

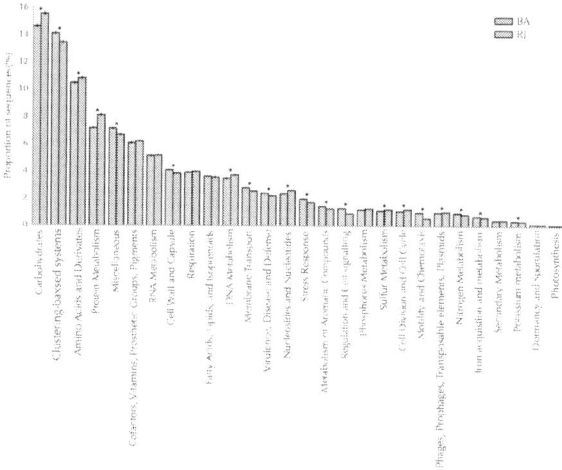

**Figure 3**: Profile bar plot showing the relative proportion of RJ (yellow) and BA (blue) sequences classified according the SEED subsystem obtained using the MG-RAST functional profile through STAMP software. The analysis was conducted considering the entire sample as the parental level and level 1 as the profile level. (*) *p*-value ≤ 0.01.

The RJ sample was found to have a significantly higher number of EGTs related to carbohydrate metabolism (Figure 3). Genes commonly found in other metagenomes related to amino acid metabolism, cellular respiration, and DNA and protein metabolism were also identified. Furthermore, sequences related to virulence, the stress response, membrane transport, cell wall and capsule synthesis, and the metabolism of aromatic compounds were also significantly more abundant in the BA mangrove. A greater number of EGTs related to nitrogen metabolism, potassium metabolism, and iron acquisition and metabolism were also identified in the BA sample. In contrast, genes involved in sulfur metabolism (Figure 3) and $CO_2$ fixation were found to be more highly represented in the RJ mangrove.

Further analyses of the functional composition of mangrove metagenomes using similarity to a non-redundant protein database against the KEGG metabolic pathways produced similar results (Figure 4A). A deep examination of the carbohydrate metabolic pathways (Figure 4B) revealed such subgroups as fructose and mannose, starch and sucrose, and galactose metabolism with a significant number of putative genes potentially involved in the decomplexation of vegetable biomass.

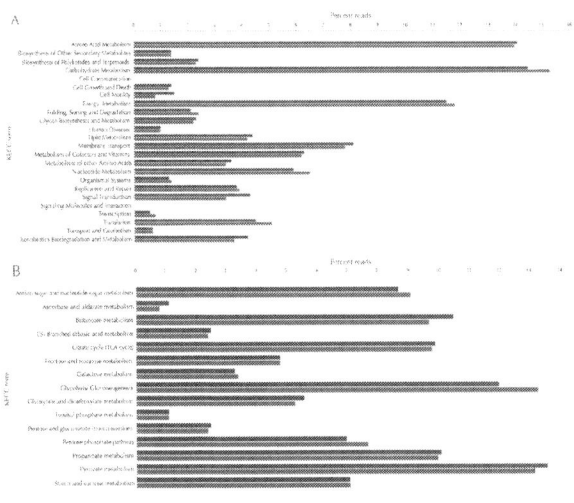

**Figure 4:** Functional analysis based on the KEGG main categories. (A) Relative number of genes from mangrove microbiomes classified according to the

KEGG main categories. (B) Relative gene diversity within the KEGG pathway for carbohydrate metabolism.

Several genes related to cellulose degradation were found, including endo-1,4-β-D-glucanases (entry 3.2.1.4; BA: 72 and RJ: 90), 1,4-β-cellobiosidases (3.2.1.91; BA: 4 and RJ: 3), and β-glucosidases (3.2.1.21; BA: 279 and RJ: 228). Sequences classified as putative xylan-degrading enzymes (3.2.1.37, 1,4-β-xylosidases; BA: 23 and RJ: 9) were also identified (Figure 5).

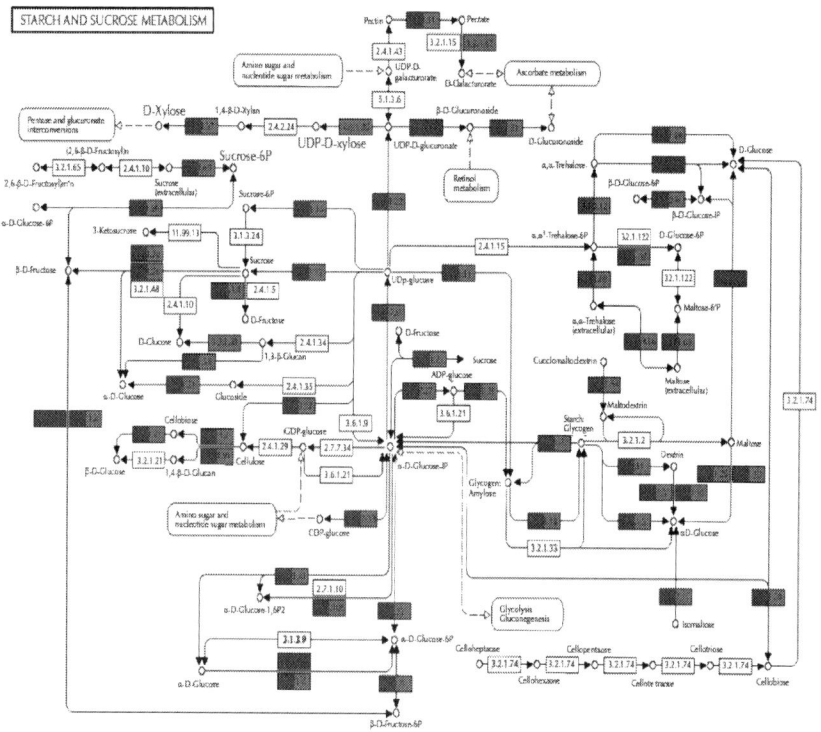

**Figure 5:** KEGG sub-pathway for starch and sucrose metabolism. The color scale represents the number of genes (in logarithmic scale) found for each KEGG entry. RJ (red) and BA (blue).

The enzymes necessary for the first step of methane metabolism, i.e., the transformation of methane into methanol, were not found. However, several genes related to the conversion of carbon dioxide into carbon monoxide and acetyl-CoA were recovered (Figure 6A). A total of 504 and 562 genes involved in the transformation of formate to $CO_2$ were identified in the BA and RJ samples, respectively. There were 75 (BA) and 103 (RJ) hydrogen dehydrogenases responsible for the interconversion of $H^+$ and $H_2$ identified in the samples.

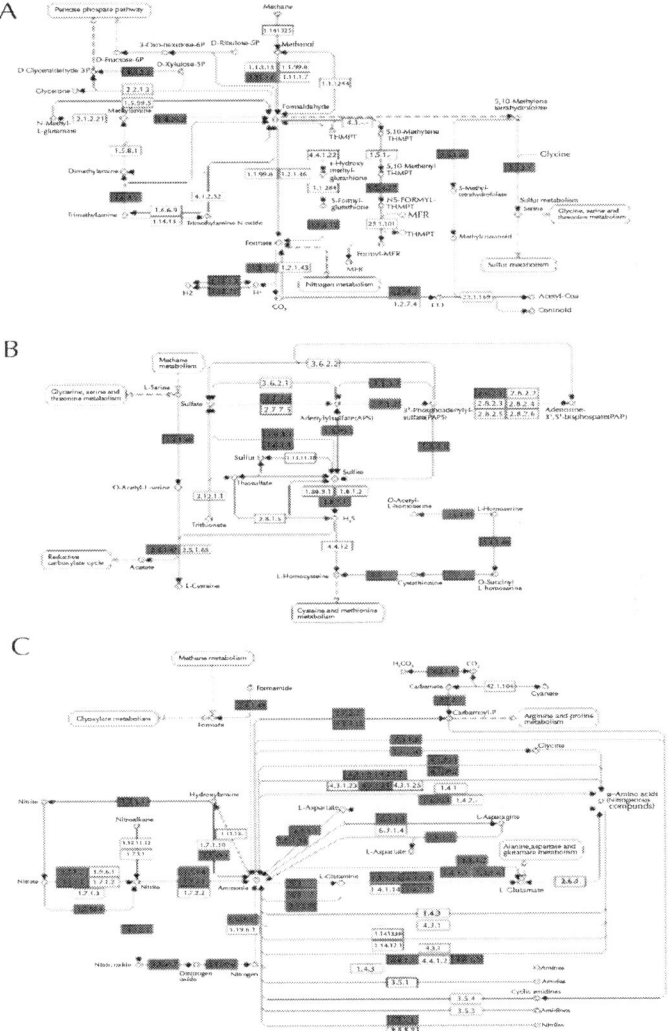

**Figure 6:** Functional analysis of the mangrove metagenomes based on the KEGG pathways. (A) methane metabolism; (B) sulfur metabolism, and (C) nitrogen metabolism.

In terms of sulfur metabolism, several enzymes in the mangroves (BA: 68 and RJ: 72) convert adenylylsulfate into sulfite (Figure 6B). The majority of enzymes involved in nitrogen metabolism are represented between the two mangroves, with the exception of methylaspartate ammonia-lyase (EC number: 4.3.1.2). This enzyme produces ammonia as a subproduct and is present only in the RJ sample (Figure 6C). The pathways leading to ammonia production are well represented by histidine ammonia-lyase (EC: 4.3.1.3), L-tryptophan indole-lyase (EC: 4.1.99.1), D-amino-acid dehydrogenase (EC: 1.4.99.1), and L-cystathionine L-homocysteine-lyase (EC: 4.4.1.8). These enzymes were found at higher abundance levels in the BA sample.

The BA and RJ samples were found to have 60 and 49 nitrogenases involved in biological nitrogen fixation, respectively. *Cyanobacteria*, which were highly abundant in both mangroves, as well as *Chlorobi* (green sulfur bacteria), *Azotobacteraceae*(*Gammaproteobacteria*), *Frankia* (*Actinobacteria*), and *Rhizobia* (*Alphaproteobacteria*) are the diazotrophs involved in this process. Enzymes related to the nitrification were thus identified. However, the denitrification process, which reduces nitrates to nitrogen gas, has an even higher number of sequences in the two mangroves. The bacteria involved in this process are anaerobic and use nitrates as an oxygen alternative to the final electron acceptor in the respiration cycle.

Metabolic pathway comparative analysis with the KEGG database indicated the presence of 202 and 183 alcohol dehydrogenases (EC number: 1.1.1.1) in BA and RJ samples, respectively, which are involved in polycyclic aromatic degradation. Additionally, 27 and 6 salicylate 1-monooxygenases from the same pathway (EC number: 1.14.13.1) were found in the BA and RJ samples, respectively. These enzymes play a role in 1- and 2- methylnaphthalene degradation.

Naphthalene is considered a PAH and is known to cause human disease (IARC, 2002). After evaluating the metabolism of xenobiotics by the cytochrome P450 pathway, two important enzymes for naphthalene metabolism were identified: glutathione dehydrogenase, which had 117 and 40 hits for BA and RJ, respectively, and benzo[a]pyrene-4,5-oxide hydratase, which had 14 and 10 hits for BA and RJ, respectively.

# Carbohydrate-Active Enzyme Analysis

Using the CAZy database (http://www.cazy.org *webcite*), the two mangrove samples were found to have the same metabolic potential to hydrolyze carbohydrates, including cellulosic compounds (cellulose, hemicellulose, pectin, xylan, among others). This analysis identified more than 2,900 Environmental Gene Tags (EGTs) from 110 different CAZy families, most of which were from *Proteobacteria*. Some CAZy families were only found in one mangrove: GH 6, 45, 70, 71, 72, and CBM 4 were only found in the RJ sample, while GH 11, 46, 79, 81, CE 5, and CBMs 15, 25, and 33 were only found in the BA sample.

A wide diversity of GH (glycosyl hydrolase) families was also identified: 54 and 56 different GH families were found in the BA and RJ mangroves, respectively. GH families with cellulolytic enzymes involved in the degradation of plant cell walls were found in both metagenomes, and only a few members of GH 6 and 45 were detected exclusively in the RJ sample. In addition, enzymes involved in the hydrolysis of hemicellulose/pectin and xylan side chains, including the β-galactosidases (GH 2, 27, 35), the xylanases (GH 10, 26, 43), the acetylxylan esterases (CE 4), the α-mannosidases (GH 38), the α-xylosidases (GH 31), the pectin methylesterases (CE 8), the α-L-rhamnosidases (GH 78), the α-glucuronidases (GH 67), and the pectin lyases (PL 1), were identified.

The RJ and BA mangroves also possess a wide diversity of carbohydrate-binding modules (CBM), which may promote the interaction between a given enzyme and its target substrate, thereby increasing its catalytic efficiency. Important modules for the degradation of cellulose were found (CBM4/9/16/22). The CBM module with the most identified members was CBM50, known as the LysM domain.

In total, 58 families of glycosyl hydrolases, 23 families of glycosyltransferases, 5 families of carbohydrate esterases, and 3 families of polysaccharide lyases were identified in both mangrove microbiomes.

When comparing mangroves to other host-associated and environmental metagenomes, including those from termites, pandas (Zhu et al. 2011), marine water systems (DeLong et al. 2006) and soil, some GHs involved in the hydrolysis of cellulosic/hemicellulosic compounds were found exclusively in the mangroves (GH 6, 12, 17, 44, 46, 47, 62, and 76). Several CBMs important in the recognition of cellulosic compounds were exclusive to the mangroves (CBM 2, 3 and 15). Families GH 3 and GH 13, which contain a large range of glycosidases able to hydrolyze complex carbohydrates into oligosaccharides, were abundant.

Crucial carbohydrate degradation-related domains, including the dockerin and cohesin domains, were found in both mangroves; the cohesins were highly represented In other metagenomes, only one dockerin (in the marine metagenome) and few cohesins (in the panda and marine metagenomes) were detected. Even in metagenomes specialized in cellulose degradation, such as the termite metagenome, no dockerin/cohesion sequences were identified, further showing the remarkable potential of the mangrove environment for enzyme identification.

Deep analyses of these results also identifies microorganism sequences specialized for cellulose/biomass degradation. Sequences belonging to *Anaerolinea thermophila*, which is a filamentous and thermophilic bacterium, were the most abundant in both mangrove samples (30 CAZy sequences). Interestingly, these results also showed the identification of CAZy sequences from *Fibrobacter succinogenes* in both microbiomes (RJ: CBM 4, 9, 16, and 22; and BA: GH 30). Sequences from other specialized microorganisms were also found, including *Sorangium cellulosum*, *Solibacter usitatus* and *Spirochaeta thermophila*.

# Comparative Metagenomic Analysis

The functional information provided by SEED was used to compare the Brazilian mangrove metagenome to the published metagenomes of other environments, including soil, water, and host-associated environments. Principal component analysis with only the first two components could explain 77.2% of the data variance. The resulting plot (Figure 7) showed that the BA and RJ mangroves formed a cluster near the metagenomes of water and soil. The metagenomes from the Caribbean Sea (open ocean), the Atlantic Ocean (4,200 m), agricultural soil, tropical forest soil, and rice rhizospheres were the most similar to the mangrove metagenome. The host-associated metagenomes formed two main groups: one formed by the human and termite sequences, and the other formed by the mouse sequences. Additionally, the metagenomes of euphotic, mesopelagic and ocean regions with a minimum oxygen layer were similar to these functional categories and formed a cluster that was separate from the others.

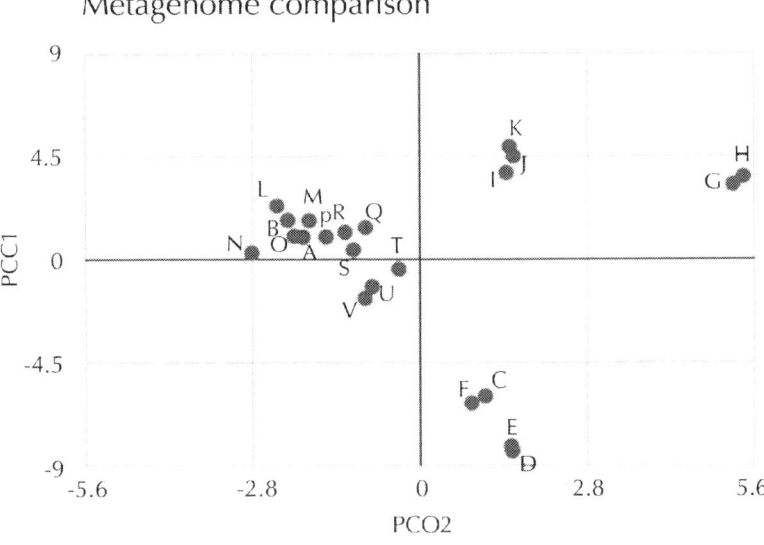

**Figure 7:** Comparative analysis of BA and RJ mangroves, water, soil, and host-associated metagenomes obtained through principal component analysis based on SEED subsystems determined using MG-RAST. The letters indicate the following metagenomes, with the respective MG-RAST acession number: (A) BA mangrove (4485982.3), (B) RJ mangrove (4485981.3), (C) Pacific Ocean, upper mesopelagic 500 m (4441057.3), (D) Pacific Ocean, upper euphotic 70 m (4441057.4), (E) Pacific Ocean, upper euphotic (4442500.3), (F) Pacific Ocean, oxygen minimum layer (4442500.4), (G)Lean mouse cecum (4440463.3), (H) Obese mouse cecum (4440464.3), (I) Human feces (4440939.3), (J) Human feces (4440941.3), (K) Termite gut (4442701.3), (L) Rice rhizosphere (4449956.3), (M) Caribbean sea - open Ocean (4441587.3), (N) Tropical forest soil (4446153.3), (O) Atlantic Ocean 4,200 m (4441572.3), (P) Agricultural soil (4441091.3), (Q) Water mangrove (4441598.3), (R) Atlantic Ocean surface (4441584.3),(S) Pacific Ocean 2,736 m (4441594.3), (T) Acid mine drainage biofilm (4441138.3), (U)Pacific Ocean coast (4443713.3), (V) Ocean coast (4443702.3).

# DISCUSSION

Mangrove ecosystems are coastal wetlands that are dominated by woody plants across a gradient of latitudes (30 °N to 37 °S), tidal heights (>1 m to <4 m), geomorphologies (oceanic islands to riverine systems), sedimentary environments (peat to alluvial), climates (warm temperate to both arid and wet tropics), and nutrient availabilities (oligotrophic to eutrophic) (Feller et al. 2010). These diverse ecosystems are critical not only for sustaining biodiversity but also for their direct and indirect benefits for human activity (Feller et al. 2010; Koch et al.2009; Walters et al. 2008). Brazil has 7,408 km of coastline, 6,786 km of which contain mangrove forests. Brazil's mangroves cover approximately 25,000 km$^2$ (Schaeffer-Novelli et al. 200).

Unfortunately, the diversity of global mangroves has been continually declining over the past four decades (Butchart et al. 2010). The world's total mangrove area has decreased by approximately 35% in the past 30 years (Giri et al. 2011). The current estimation of mangrove forest areas in the world is less than half of what it once was (Spiers et al. 1999; Spalding et al.1997) and most of the remaining areas are already in a degraded state (Giri et al. 2011). Predictions suggest that a full 100% of the mangrove forests, along with their vast microbial diversity and intrinsic benefits, may be lost within the next 100 years, given the accelerated loss due to both human encroachment and environmental changes (Duke et al. 2007).

Our results are consistent with these predictions; the impacted BA mangrove was less diverse than the well-preserved RJ mangrove. The estimated disappearance rates for mangrove areas vary from between 1 and 8% per year, which is as high as or even higher than the disappearance rates for tropical wet forests. These rates will likely continue to increase, unless mangrove forests can become protected as a valuable resource (Alongi et al. 2002). The exploration of the biotechnological potential of mangroves worldwide, as a source for the discovery of new or improved enzymes and microorganisms for second-generation biofuel production, would have several simultaneous benefits. It could result in the increased protection and preservation of these ecosystems, leading to the maintenance of their ecological and biotechnological potential. It will therefore be useful to optimize and increase the production and consumption of biofuels based on agricultural residues and to thereby decrease the consumption of fossil fuels.

Given their role in vegetable biomass degradation, the neglected mangrove ecosystems have significant biotechnological potential and represent an excellent microbiome for the study of cellulolytic microorganisms and biomass degrading enzymes. The unique characteristics of mangroves include their differences in salinity, oxygen content and nutrients, which are responsible for the high level of diversity and exclusivity of the microorganisms (Gao et al.2010).

A main goal of modern biotechnological approaches is to access the microbes from unusual or specialized ecological niches, such as mangrove habitats, the rumens of herbivorous animals (Hess et al. 2011; Zhu et al. 2011; Brulc et al. 2009) and the guts of arthropods (Warnecke et al. 2007). All of these microbiomes have the metabolic potential to hydrolyze cellulolytic compounds. A clear difference between searching for enzymes and microorganisms in the mangroves versus in host-associated environments is that the latter represent "closed" environments, with minimal fluctuations in pH, temperature and other environmental parameters that might thus produce a more limited level microbial diversity. The greater microbial diversity of the mangroves and their associated microbial communities may be due to the numerous daily stresses that these organisms face (Xu 2011; Pointing et al. 1999).

Certain microorganisms have developed unique metabolic pathways, ecological adaptability, defenses, ability to communicate, and degradation and predation functions. These microorganisms might therefore produce and secrete effective molecules or enzymes with promising biotechnological applications (Xu 2011). Mangroves sequester up to 25.5 million tonnes of carbon per year (Ong 1993) and provide more than 10% of the essential organic carbon to the global oceans (Dittmar et al. 2006) as such, they are considered an environment of intense carbon flux. Additionally, the wealth and diversity of the microbial community associated with vegetal decomposition (Sahoo and Dhal 2009) makes this unique ecological habitat a potential source of microorganisms and enzymes involved in cellulose decomplexation and sugar release.

Other groups have described the results of mangrove sequencing (Gomes et al. 2008; Andreote et al. 2012; Santos et al. 2011; Dias et al. 2010) and have used isolation as a strategy to assess the biodiversity and enzymatic potential of mangrove microorganisms for their use in biotechnological purposes. In this work, microorganism cultivation and a metagenomic approach were used to explore the biodiversity and biotechnological potential of the mangroves.

According to Gao et al. (2010), microorganisms from extreme environments and their unique enzymatic repertoire have a great potential value for use in biotechnology. The mangrove ecosystem is not generally considered an extreme environment, but its unique environmental features make it a similarly attractive system to look for microorganisms with useful applications. In a recent review, Xu (2011) describes several natural products that can be extracted from mangrove-associated microorganisms, with a particular focus on their bioactivities. Although the biotechnological potential of mangrove microorganisms has been an underrepresented research topic, two cellulase genes were identified in the bacterial isolates of the mangrove soils and were subsequently characterized (Gao et al. 2010; Yang et al. 2001), demonstrating the potential of this environment to degrade cellulosic compounds.

In our study, 179 (60.5%) and 117 (39.5%) microorganisms (bacteria, yeast and filamentous fungi) were isolated from the RJ and BA samples, respectively (Figure 1A), including 21 filamentous fungi with cellulolytic potential. Sahoo and Dhal (2009) reviewed the microbial diversity of mangrove ecosystems and highlighted the crucial role of these fungi as the primary microorganisms responsible for the decomposition of cellulosic material. These organisms can tolerate the high levels of phenolic compounds found on mangrove leaves that commonly inhibit the growth of other microorganisms (Raghukumar et al. 1995). They are therefore able to begin the process of vegetative material decomposition and permit the secondary colonization of bacteria and yeasts, which can then further decompose the organic matter (Sahoo and Dhal 2009). However, little is known about the physiology and biochemistry of mangrove fungi. This lack of knowledge could limit the isolation and study of the role of such microorganisms in nutrient recycling (Sahoo and Dhal 2009).

When considering the microbial diversity as a whole, the alpha diversity indicated that the RJ mangrove (alpha-diversity = 974.385 species) is more diverse than the BA mangrove (alpha-diversity = 847.721

species). This index could explain the higher number of microbial isolates in the RJ mangrove compared to the BA mangrove (Figure 1A). This result is also consistent with the conservation states of RJ and BA (RJ is highly preserved, whereas BA has a low to medium level of degradation) because the degradation of mangrove areas is closely related a decrease in their biodiversity levels (Giri et al. 2011).

Considering the different taxonomic groups found on the mangroves using metagenomic approaches, the high level of diversity found in the mangrove samples is expected because multiple species may perform both different and similar roles (Hooper er al. 2005). They can thus provide stability to the mangrove ecosystem during times of environmental disturbance (Gomes et al. 2008). The identified predominance of *Gammaproteobacteria* is consistent with previous data from Andreote et al. (2012) and Santos et al. (2011), indicating the dominance of this group in both natural and impacted environments. Dias et al. (2010) reported that both dominant and low-density communities are influenced by environmental changes, indicating that these groups are essential in the maintenance of ecosystem functionality during daily or yearly changes in climate.

Sequences belonging to *Deltaproteobacteria*, which have not been commonly found in seawater and soil samples, were identified in the preserved RJ mangrove ecosystem. It may be that the anaerobic conditions of the RJ mangrove provide a selective pressure leading to the prevalence of microbial groups such as sulfate-reducing bacteria (Taketani et al. 2010a; Taketani et al. 2010b).

In relation to the general results related to the functional aspects of the mangroves, the RJ sample presented more genes involved in sulfur metabolism (Figure 3) and $CO_2$ fixation. The biogeochemical cycles acting on the BA and RJ mangroves were evaluated, particularly in regards to their methane, sulfur and nitrogen metabolism. Although these environments are predominantly anoxic, sea flooding produces occasional aerobic conditions, thereby providing an opportunity for the nitrification process to occur. Enzymes involved in the direct conversion of sulfite into $H_2S$ through sulfite reductase were only identified in the RJ mangrove. It is important to note that the RJ sample had a higher number of *Deltaproteobacteria*, which are related to sulfate reduction. This finding is evidence for the importance of sulfur metabolism in the RJ mangrove.

A higher number of enzymes involved in the polycyclic aromatic degradation were found in the BA sample. Due to their high primary productivity, abundant detritus, rich organic matter and anoxic/reduced conditions, the mangroves are preferential sites for the uptake and preservation of PAHs, resultant of high anthropogenic impact (Bernard et al. 1996). The polycyclic aromatic compounds have several possible mechanisms for environmental release, including volatilization, photo-oxidation, chemical oxidation, bioaccumulation, adsorption on soil particles, leaching, and microbial degradation. Importantly, microbial degradation is believed to be the most important process for the successful removal of such compounds. The microorganisms of hydrocarbon-contaminated areas have a higher biodegradation potential than those from non-contaminated environments, especially in terms of their acclimation function and their adaptation to contaminated areas (Yun et al. 2008). The atmospheric pollution of the urban area of Porto Seguro may be the reason why we found a higher number of enzymes involved in the degradation of polycyclic aromatic compounds in the BA mangrove. Indeed, these contaminants are spreading along the coast where the BA mangroves are found. Peixoto et al. (2011) have demonstrated that the total level of PAHs is related to a higher abundance of Actinobacteria and Alphaproteobacteria in the samples, which is in agreement with the results for BA mangrove. Additionally, a recent study (Arias et al., 2010) showed that the proximity to sources was the most important determining factor for the distribution of PAHs, with the higher concentrations being greater in samples collected near industrial areas.

Several cellulolytic enzymes were found on the mangroves, some of them being only found in one mangrove. This finding indicates that exploring different mangrove areas and samples simultaneously might provide a higher coverage of the different enzyme families. Given that Brazil contains 7% of the world's total mangrove area (Giri et al. 2011) spread across various regions with distinct environmental conditions, the potential for discovery is remarkable. The mangrove environment should be considered a good source of carbohydrate-degrading enzymes.

The enormous GH variability found highlights the great potential of this environment for the identification of polysaccharide and cellulosic-degrading enzymes. These enzymes are very important in biomass decomplexation for biofuel production because they open the cellulosic fibers. The abundance of the GH and CBM families identified in the mangroves reflects their high capacity to attach to, to metabolize and to degrade a diverse array of cellulosic substrates. In contrast, only one family likely to bind to cellulose was found in the bovine rumen (Brulc et al. 2009).

The GH family's 3 and 13, which were found on the mangroves, were also found to be abundant in previously reported metagenomes (Pope et al. 2010; Brulc et al. 2009; Warnecke et al. 2007). The GH 3 family is known to have a bi-functional mechanism of action and has been reported to have cellobiase activity (Faure et al. 2011). The specificity of cohesin-dockerin interactions is critically important for the assembly of the multienzyme cellulolytic complex (Slutzki et al. 2012). This so-called 'cellulosome' is defined as a multienzyme and is highly active against crystalline cellulose and related plant cell wall polysaccharides (Cha et al. 2007). Each cohesin domain consists of a subunit-binding domain that interacts with a docking domain (dockerin) for each catalytic component of the cellulosomal enzyme (Cha et al. 2007). The efficient enzymatic degradation of insoluble polysaccharides requires both the tight interaction between the enzymes and their substrates and the cooperation of multiple enzymes to enhance the hydrolysis, leading to the formation of a complex structure (Slutzki et al. 2012; Cha et al. 2007; Patthra et al. 2006; Takagi et al.1993). The attachment of the cellulosome to its substrate is mediated by a cellulose-binding module (CBM) that comprises part of the cellulosome subunit (Fontes et al. 2010; Doi et al.2004). This structure was widely found in the two mangrove samples. Therefore, all of the necessary molecular machinery for cellulosome production, efficient cellulose degradation, and sugar release were found to be present in the mangrove ecosystem.

The CBM module with the most identified members was CBM50, known as the LysM domain. This domain has multiple functions, including signaling and recognition in host-microbe interactions. It is involved in symbiotic and cell wall degradation (Bensmihen et al.2011).

Among the species identified in our study is *A. thermophila*, which could be a good candidate for further studies because there are only a few reports about it. Moreover, it is known that this bacterium can produce an increase in the carbohydrate concentrations when growing in the up flow of an anaerobic sludge blanket reactor (Sekiguchi et al. 2003). According to these results,*A. thermophila* has a promising but yet-unknown cellulolytic potential that should be explored in future work. *Fibrobacter succinogenes* found in both microbiomes was prominent in the rumen of herbivores and is considered to be specialized in using only cellulose as its carbon source (Suen et al. 2010). In this case, polysaccharide-degrading strategy is different from that of other cellulolytic microorganisms, as it does not possess a vast repertoire of cellulases or cellulosomal structures. *F. succinogenes* adheres to solid cellulosic substrate, most likely forming a biofilm on the cellulose surface (Suen et al. 2010). Sequences from other specialized microorganisms were also found, including *Sorangium cellulosum*, a cellulolytic myxobacterium that can efficiently degrade many types of polysaccharides such as cellulose (Wang et al. 2007), *Solibacter usitatus*, an *Acidobacteria* found in soils and sediments worldwide, with four times more genes for carbohydrate metabolism and transport than other known *Acidobacteria* (Challacombe et al. 2011) and *Spirochaeta thermophila*, a thermophilic, free-living, and cellulolytic anaerobe, found only in the RJ mangrove (Angelov et al. 2011).

The biodiversity and biotechnological potential of two Brazilian mangroves were assessed through cultivation and metagenomic approaches. The results presented here demonstrate the great biotechnological potential of this unique but endangered environment.

# AUTHORS' CONTRIBUTIONS

Performed the experiments: MB; Analyzed the experimental data: WOB, LS; Analyzed the statistical, taxonomic, and comparative data: CET; Analyzed the functional data: CET, LS, WOBS; Contributed for the reagents/materials: ATRV, JAG, MHV, and WOBS; Wrote the manuscript: CET, LS, and WOBS; Participated in the design and coordination of the study: ATRV, JAG, MHV. All authors read and approved the final manuscript.

# ACKNOWLEDGMENTS

We would like to thank the staff of LNCC, Eloi Garcia (Fiocruz/ INMETRO), and Wanderley de Souza (UFRJ/INMETRO) for insightful discussions, to Janaína Cavalcanti (INMETRO) for the DNA preparation, and the editors from American Journal Experts (AJE) for professional language editing services. This work was supported by Fundação Carlos Chagas Filho de Amparo à Pesquisa do Estado do Rio de Janeiro (FAPERJ), Coordenação de Aperfeiçoamento de Pessoal de Nível Superior (CAPES) and Conselho Nacional de Desenvolvimento Científico e Tecnológico (CNPq).

# REFERENCES

1.    Almeida LG, Paixão R, Souza RC, da Costa GC, Barrientos FJA, dos Santos MT, de Almeida D, Vasconcelos ATR (2004) A System for Automated Bacterial (genome) Integrated Annotation -SABIA. Bioinformatics 20:2832-2833

2.    Alongi DM (2002) Present state and future of the world's mangrove forests. Environ Conserv 29:331-349

3.    Altschul SF, Madden TL, Schäffer AA, Zhang J, Zhang Z, Miller W, Lipman DJ (1997) Gapped BLAST and PSI-BLAST (1997) a new generation of protein database search programs. Nucleic Acids Res 25:3389-3402

4.    Andreote FD, Jiménez DJ, Chaves D, Dias ACF, Luvizotto DM, Dini-Andreote Fasanella CC, Lopez MV, Baena S, Taketani RG, de Melo IS (2012) The microbiome of brazilian mangrove sediments as revealed by metagenomics. PLOS ONE 7:e38600

5.    Angelov A, Loderer C, Pompei S, Liebl W (2011) Novel family of carbohydrate-binding modules revealed by the genome sequence of *Spirochaeta thermophila* DSM 6192. Appl Environ Microbiol 77(15):5483-5489

6.    Arias AH, Vazquez-Botello A, Tombesi N, Ponce-Vélez G, Freije H, Marcovecchio J (2010) Presence, distribution, and origins of polycyclic aromatic hydrocarbons (PAHs) in sediments from Bahía Blanca estuary, Argentina. Environ Monit Assess 160:301-314

7.  Bensmihen S, de Billy F, Gough C (2011) Contribution of NFP LysM domains to the recognition of Nod factors during the *Medicago truncatula/Sinorhizobium meliloti* symbiosis. PLOS ONE 6(11):e26114

8.  Bernard D, Pascaline H, Jeremie JJ (1996) Distribution and origin of hydrocarbons in sediments from lagoons with fringing mangrove communities. Mar Pollut Bull 32:734-739

9.  Brulc JM, Antonopoulos DA, Miller MEB, Wilson MK, Yannarell AC, Dinsdale EA, Edwards RE, Frank ED, Emerson JB, Wacklin P, Coutinho PM, Henrissat B, Nelson KE, White BA (2009) Gene-centric metagenomics of the fiber-adherent bovine rumen microbiome reveals forage specific glycoside hydrolases. Proc Natl Acad Sci U S A 106:1948-1953

10. Butchart SHM, Walpole M, Collen B, van Strien A, Scharlemann JPW, Almond REA, Baillie JEM, Bomhard B, Brown C, Bruno J, Carpenter KE, Carr GM, Chanson J, Chenery AM, Csirke J, Davidson NC, Dentener F, Foster M, Galli A, Galloway JN, Genovesi P, Gregory RD, Hockings M, Kapos V, Lamarque J-F, Leverington F, Loh J, McGeoch MA, McRae L, Minasyan A, et al. (2010) Global biodiversity: indicators of recent declines. Science 328:1164-1168

11. Cantarel BL, Coutinho PM, Rancurel C, Bernard T, Lombard V, Henrissat B (2009) The Carbohydrate-Active EnZymes database (CAZy): an expert resource for glycogenomics. Nucleic Acids Res 37:233-238

12. Cha J, Matsuoka S, Chan H, Yukawa H, Inui M, Roy H (2007) Effect of multiple copies of cohesins on cellulase and hemicellulase activities of Clostridium cellulovorans mini-cellulosomes. J Microbiol Biotechnol 17(11):1782-1788

13. Challacombe JF, Eichorst SA, Hauser L, Land M, Xie G, Kuske CR (2011) Biological consequences of ancient gene acquisition and duplication in the large genome of *Candidatus Solibacter usitatus* Ellin6076. PLOS ONE 6(9):e24882

14. Chou HH, Holmes MH (2001) DNA sequence quality trimming and vector removal. Bioinformatics 17:1093-1104

15. Cole JR, Wang Q, Cardenas E, Fish J, Chai B, Farris RJ, Kulam-Syed-Mohideen AS, McGarrell DM, Marsh TL, Garrity GM, Tiedje

JM (2009) The Ribosomal Database Project: improved alignments and new tools for rRNA analysis. Nucleic Acids Res 37:141-145

16. DeLong EF, Preston CM, Mincer T, Rich V, Hallam SJ, Frigaard N-U, Martinez A, Sullivan MB, Edwards R, Brito BR, Chisholm SW, Karl DM (2006) Community genomics among stratified microbial assemblages in the ocean's interior. Science 311:496-503

17. Dias ACF, Andreote FD, Dini-Andreote F, Lacava PT, Sá ALB, Melo IS, Azevedo JL, Araújo WL (2009) Diversity and biotechnological potential of culturable bacteria from Brazilian mangrove sediment. World J Microb Biot 25(7):1305-1311

18. Dias ACF, Andreote FD, Rigonato J, Fiore MF, Melo IS, Araujo W (2010) The bacterial diversity in a Brazilian non-disturbed mangrove sediment. Antoine van Leeuwenhoek 98:541-551

19. Dittmar T, Hertkorn N, Kattner G, Lara RJ (2006) Mangroves, a major source of dissolved organic carbon to the oceans. Global Biogeochem Cy 20:GB1012-GB1019

20. Doi RH, Kosugi A (2004) Cellulosomes: plant-cell-wall-degrading enzyme complexes. Nat Rev Microbiol 2(7):541-551

21. Dourado MN, Ferreira A, Araújo WL, Azevedo JL, Lacava PT (2012) The diversity of endophytic methylotrophic bacteria in an oil-contaminated and an oil-free mangrove ecosystem and their tolerance to heavy metals. Biotechnol Res Int 2012:1-8

22. Duan CJ, Feng JX (2010) Mining metagenomes for novel cellulose genes. Biotechnol Lett 32:1765-1775

23. Duke NC, Meynecke JO, Dittmann S, Ellison AM, Anger K, Berger U, Cannici S, Diele K, Ewel KC, Field CD, Koedam N, Lee SY, Marchand C, Nordhaus I, Dahdouh-Guebas F (2007) A world without mangroves? Science 317:41-42

24. Faure D, Henrissat B, Ptacek D, Bekri MA, Vanderleyden J (2011) The celA gene, encoding a glycosyl hydrolase family 3 β-glucosidase in Azospirillum irakense is required for optimal growth on cellobiosides. Appl Environ Microbiol 67(5):2380-2383

25. Feller IC, Lovelock CE, Berger U, McKee KL, Joye SB, Ball MC (2010) Biocomplexity in mangrove ecosystems. Ann Rev Mar Sci 2:395-417

26.  Finn RD, Clements J, Eddy SR (2011) HMMER web server: interactive sequence similarity searching. Nucleic Acids Res 39:29-37

27.  Fontes CM, Gilbert HJ (2010) Cellulosomes: highly efficient nanomachines designed to deconstruct plant cell wall complex carbohydrates. Annu Rev Biochem. 79:655-681

28.  Gao Z, Ruan L, Chen X, Zhang Y, Xu X (2010) A novel salt-tolerant endo-β-1,4-glucanase Cel5A in *Vibrio* sp. G21 isolated from mangrove soil. Appl Microbiol Biotechnol 87:1373-1382

29.  Giri C, Ochieng E, Tieszen LL, Singh A, Loveland T, Masek J, Duke N (2011)) Status and distribution of mangrove forests of the world using earth observation satellite data. Global Ecol Biogeogr 20:154-159

30.  Gomes NCM, Borges LR, Paranhos R, Pinto FN, Mendonça-Hagler LCS, Smalla K (2008) Exporing the diversity of bacterial communities in sediments of urban mangrove forests. FEMS Microbiol Ecology 2008(66):96-109

31.  Gomez-Alvarez V, Teal TK, Schmidt TM (2009) Systematic artifacts in metagenomes from complex microbial communities. ISME J 3:1314-1317

32.  Gonzales-Acosta B, Bashan Y, Hernandez-Saavedra NY, Ascencio F, De la Cruz-Aguero G (2006) Seasonal seawater temperature as the major determinant for populations of culturable bacteria in the sediments of an intact mangrove in an arid region. FEMS Microbiol Ecol 55:311-321

33.  Hess M, Sczyrba A, Egan R, Kim TW, Chokhawala H, Schroth G, Luo S, Clark DS, Chen F, Zhang T (2011) Metagenomic discovery of biomass-degrading genes and genomes from cow rumen. Science 331:463-467

34.  Holguin G, Vazquez P, Bashan Y (2001) The role of sediment microorganisms in the productivity, conservation, and rehabilitation of mangrove ecosystems: an overview. Biol Fert Soils 33(4):265-278

35.  Hooper DU, Chapin FS III, Ewel JJ, Hector A, Inchausti P, Lavorel S, Lawton JH, Lodge DM, Loreau M, Naeem S, Schmid B, Setälä H, Symstad AJ, Vandermeer J, Wardle DA (2005) Effects of biodiversity on ecosystem functioning: a consensus of current knowledge. Ecol Monogr 75:3-35

36. Huang Y, Gilna P, Li W (2009) Identification of ribosomal RNA genes in metagenomic fragments. Bioinformatics 25:1338-1340

37. Huson DH, Mitra S, Ruscheweyh HJ, Weber N, Schuster SC (2011) Integrative analysis of environmental sequences using MEGAN4. Genome Res 21:1552-1560

38. International Agency for Research on Cancer (IARC) (2002) Traditional Herbal Medicines, Some Mycotoxins, Naphthalene and Styrene. IARC Monographs on the Evaluation of Carcinogenic Risks to Humans, vol 82. Lyon, France: IARC.

39. Kanehisa M, Goto S, Sato Y, Furumichi M, Tanabe M (2012) KEGG for integration and interpretation of large-scale molecular data sets. Nucleic Acids Res 40:D109-D114

40. Koch EW, Barbier EB, Silliman BR, Reed DJ, Perillo GME, Hacker SD, Granek EF, Primavera JH, Muthiga N, Polasky S, Halpern BS, Kennedy CJ, Kappel CV, Wolanski E (2009) Nonlinearity in ecosystem services: temporal and spatial variability in coastal protection. Front Ecol Environ 7:29-37

41. Kumar R, Singht S, Singh OV (2008) Bioconversion of lignocellulosic biomass: biochemical and molecular perspectives. J Ind Microbiol Biotechnol 35:377-391

42. Medie FM, Davies GJ, Drancourt M, Henrissat B (2012) Genome analyses highlight the different biological roles of cellulases. Nature Rev Microbiol 10:227-234

43. Meyer F, Paarmann D, D'Souza M, Olson R, Glass EM, Kubal M, Paczian T, Rodriguez A, Stevens R, Wilke A, Wilkening J, Edwards RA (2008) The metagenomics RAST server - a public resource for the automatic phylogenetic and functional analysis of metagenomes. BMC Bioinformatics 9:386-394

44. Mitra S, Klar B, Huson DH (2009) Visual and statistical comparison of metagenomes. Bioinform 25:1849-1855

45. Ong JE (1993) Mangroves – a carbon source and sink. Chemosphere 27:1097-1107

46. Overbeek R, Begley T, Butler RM, Choudhuri JV, Chuang H-Y, Cohoon M, Crécy-Lagard V, Diaz N, Disz T, Edwards R, Fonstein M, Frank ED, Gerdes S, Glass EM, Goesmann A, Hanson A, Iwata-Reuyl D, Jensen R, Jamshidi N, Krause L, Kubal M, Larsen N, Linke B, Mc-Hardy AC, Meyer F, Neuweger H, Olsen G, Olson

R, Osterman A, Portnoy V, et al. (2005) The subsystems approach to genome annotation and its use in the project to annotate 1000 genomes. Nucleic Acids Res 33:5691-5702

47. Parks DH, Beiko RG (2010) Identifying biologically relevant differences between metagenomic communities. Bioinformatics 26:715-721

48. Patthra P, Chon GH, Ratanakhanokchai K, Kyu KL, Jhee O-H, Juseop K, Kim WH, Kyung-Min C, Gil-Soon P, Jin-Sang L, Hyun P, Rho MS, Yun-Sik L (2006) Selection of multienzyme complex-producing bacteria under aerobic cultivation. J Microbiol Biotechnol 16:1269-1275

49. Peixoto R, Chaer GM, Carmo FL, Araújo FV, Paes JE, Volpon A, Santiago GA, Rosado AS (2011) Bacterial communities reflect the spatial variation in pollutant levels in Brazilian mangrove sediment. Antonie van Leeuwenhoek 99:341-354

50. Pointing SB, Buswell JA, Jones EBG, Vrijmoed LLP (1999) Extracellular cellulolytic enzyme profiles of five lignicolous mangrove fungi. Mycol Res 103:696-700

51. Pope PB, Denman SE, Jones M, Tringe SG, Barry K, Malfatti SA, McHardy AC, Cheng JF, Hugenholtz P, McSweeney CS, Morrison M (2010) Adaptation to herbivory by the Tammar wallaby includes bacterial and glycoside hydrolase profiles different from other herbivores. Proc Natl Acad Sci USA 107(33):14793-14798

52. Punta M, Coggill PC, Eberhardt RY, Mistry J, Tate J, Boursnell C, Pang N, Forslund K, Ceric G, Clements J, Heger A, Holm L, Sonnhammer ELL, Eddy SR, Bateman A, Finn RD (2012) The Pfam protein families database. Nucleic Acids Res 40:290-301

53. Raghukumar S, Sathe-Pathak V, Sharma S, Raghukumar C (1995) Thraustochytrid and fungal component of marine detritus. III. Field studies on decomposition of leaves of the mangrove*Rhizophora apicula*. Aquatic Microbial Ecology 9:117-125

54. Rho M, Tang H, Ye Y (2010) FragGeneScan: predicting genes in short and error-prone reads. Nucleic Acids Res 38:e121

55. Sahoo K, Dhal NK (2009) Potential microbial diversity in mangrove ecosystems: a review. Indian J Mar Sci 38:249-256

56. Santos H, Cury JC, Carmo FL, Santos AL, Tiedje J, van Elsas JD, Rosado AS, Peixoto RS (2011) Mangrove bacterial diversity and

the impact of oil contamination revealed by pyrosequencing: bacterial proxies for oil pollution. PLOS ONE 6:e16943

57. Sekiguchi Y, Yamada T, Hanada S, Ohashi A, Harada H, Kamagata Y (2003) *Anaerolinea thermophila* gen. nov., sp. nov. and *Caldilinea aerophila* gen. nov., sp. nov., novel filamentous thermophiles that represent a previously uncultured lineage of the domain Bacteria at the subphylum level. Int J Syst Evol Microbiol 53(Pt 6):1843-1851

58. Slutzki M, Ruimy V, Morag E, Barak Y, Haimovitz R, Lamed R, Bayer EA (2012) High-throughput screening of cohesin mutant libraries on cellulose microarrays. Methods Enzymol 510:453-463

59. Spalding MD, Blasco F, Field CD (eds) (1997) World mangrove atlas Okinawa, Japan: The International Society for Mangrove Ecosystems.

60. Spiers AG (1999)) Review of international/continental wetland resources. Global review of wetland resources and priorities for wetland inventory. In: Finlayson CM, Spiers AG (eds) Supervising Scientist Report 144, Canberra, Australia: Australian Government, Department of The Environment - Supervising Scientist Division. pp 63-104

61. Suen G, Scott JJ, Aylward FO, Adams SM, Tringe SG, Pinto-Tomás AA, Foster CE, Pauly M, Weimer PJ, Barry KW, Goodwin LA, Bouffard P, Li L, Osterberger J, Harkins TT, Slater SC, Donohue TJ, Currie CR (2010) An insect herbivore microbiome with high plant biomass-degrading capacity. PLoS Genet 6:e1001129

62. Sun FL, Wang YS, Sun CC, Peng YL, Deng C (2012) Effects of three different PAHs on nitrogen-fixing bacterial diversity in mangrove sediment. Ecotoxicol 21:1651-1660

63. Takagi M, Hashida S, Goldstein MA, Doi RH (1993) The hydrophobic repeated domain of the *Clostridium cellulovorans* cellulose-binding protein (CbpA) has specific interactions with endoglucanases. J Bacteriol 175:7119-7122

64. Taketani RG, Franco NO, Rosado AS, van Elsas JD (2010a) Microbial community response to a simulated hydrocarbon spill in mangrove sediments. J Microbiol 48(1):7-15

65. Taketani RG, Yoshiura CA, Dias ACF, Andreote FD, Tsai SM (2010b) Diversity and identification of methanogenic archaea

and sulphate-reducing bacteria in sediments from a pristine tropical mangrove. Antonie van Leeuwenhoek 97(4):401-411

66.  Teather RM, Wood PJ (1982) Use of Congo red-polysaccharide interactions in enumeration and characterization of cellulolytic bacteria from the bovine rumen. Appl Environ Microbiol 43(4):777-780

67.  Tringe SG, von Mering C, Kobayashi A, Salamov AA, Chen K, Chang HW, Podar M, Short JM, Mathur EJ, Detter JC, Bork P, Hugenholtz P, Rubin EM (2005) Comparative metagenomics of microbial communities. Science 308:554-557

68.  Walters BB, Ronnback P, Knovacs JM, Crona B, Hussain SA, Badola R, Primavera JH, Barbier E, Dahdouh-Guebas F (2008) Ethobiology, socio-economics and management of mangrove forests: a review. Aquat Bot 89:220-236

69.  Wang Q, Garrity GM, Tiedje JM, Cole JR (2007) Naive bayesian classifier for rapid assignment of rRNA sequences into the new bacterial taxonomy. Appl Environ Microbiol 73:5261-5267

70.  Warnecke F, Luginbühl P, Ivanova N, Ghassemian M, Richardson TH, Stege JT, Cayouette M, McHardy AC, Djordjevic G, Aboushadi N, Sorek R, Tringe SG, Podar M, Martin HG, Kunin V, Dalevi D, Madejska J, Kirton E, Platt D, Szeto E, Salamov A, Barry K, Mikhailova N, Kyrpides NC, Matson EG, Ottesen EA, Zhang X, Hernándes M, Murillo C, Acosta LG, et al. (2007) Metagenomic and functional analysis of hindgut micro-biota of a wood-feeding higher termite. Nature 450:560-565

71.  Willner D, Thurber RV, Rohwer F (2009) Metagenomic signatures of 86 microbial and viral metagenomes. Environ Microbiol 11:1752-1766

72.  Xu J (2011) Biomolecules produced by mangrove-associated microbes. Curr Med Chem 18:5224-5266

73.  Yang ZY, Chen RZ, Yang F, Xu X (2001) Cloning and DNA sequencing of Bacillus pumilusendo-1, 4-β-glucanase gene. Wei Sheng Wu Xue Bao 41:76-81

74.  Yun T, Yuan-rong L, Tian-ling Z, Li-xhe C, Xiao-xing C, Chong-ling Y (2008) Contamination and potential biodegradation of polycyclic aromatic hydrocarbons in mangrove sediments of Xiamen, China. Mar Pollut Bull 56:1184-1191

75.  Zhou HW, Wong AHY, Yu RMK, Park YD, Wong YS, Tam NFY (2008) Polycyclic aromatic hydrocarbon-induced structural shift of bacterial communities in mangrove sediment. Microbiol Ecol 58:153-160

76.  Zhu W, Lomsadze A, Borodovsky M (2010) *Ab initio* gene identification in metagenomic sequences. Nucleic Acids Res 38:e132

77.  Zhu L, Wu Q, Dai J, Zhang S, Wei F (2011) Evidence of cellulose metabolism by the giant panda gut microbiome. Proc Natl Acad Sci USA 108:17714-17719

# Chapter 4

# Current Progress on Bio-Based Polymers and Their Future Trends

Ramesh P Babu[1,2,] Kevin O'Connor[3,] and Ramakrishna Seeram[4,5,6]

[1]Centre for Research Adoptive Nanostructures and Nano Devices, Trinity College, Dublin 2, Ireland

[2]School of Physics, Trinity College Dublin, Dublin 2, Ireland

[3]School of Biomolecular and Biomedical Sciences, Centre for Synthesis and Chemical Biology, UCD Conway Institute, and Earth Institute, University College Dublin, Belfield, Dublin 4, Ireland

[4]NUSNNI, National University of Singapore, 2 Engineering Drive 3, Singapore, 117581, Singapore

[5]Institute of Materials Research and Engineering, Singapore, 117602, Singapore

[6]Jinan University, Guangzhou, China

# ABSTRACT

This article reviews the recent trends, developments, and future applications of bio-based polymers produced from renewable resources. Bio-based polymers are attracting increased attention due to environmental concerns and the realization that global petroleum resources are finite. Bio-based polymers not only replace existing polymers in a number of applications but also provide new combinations of properties for new applications. A range of bio-based polymers are presented in this review, focusing on general methods of production, properties, and commercial applications. The review examines the technological and future challenges discussed in bringing these materials to a wide range of applications, together with potential solutions, as well as discusses the major industry players who are bringing these materials to the market.

# REVIEW

## Introduction

Bio-based polymers are materials which are produced from renewable resources. The terms bio-based polymers and biodegradable polymers are used extensively in the literature, but there is a key difference between the two types of polymers. Biodegradable polymers are defined as materials whose physical and chemical properties undergo deterioration and completely degrade when exposed to microorganisms, carbon dioxide (aerobic) processes, methane (anaerobic processes), and water (aerobic and anaerobic processes). Bio-based polymers can be biodegradable (e.g., polylactic acid) or nondegradable (e.g., biopolyhethylene). Similarly, while many bio-based polymers are biodegradable (e.g., starch and polyhydroxyalkanoates), not all biodegradable polymers are bio-based (e.g., polycaprolactone).

Bio-based polymers still hold a tiny fraction of the total global plastic market. Currently, biopolymers share less than 1% of the total market. At the current growth rate, it is expected that biopolymers will account for just over 1% of polymers by 2015 (Doug 2010).

The worldwide interest in bio-based polymers has accelerated in recent years due to the desire and need to find non-fossil fuel-based polymers. As indicated by ISI Web of Sciences and Thomas Innovations, there is a tremendous increase in the number of publication citations on bio-based polymers and applications in recent years, as shown in Figure 1 (Chen and Martin 2012)

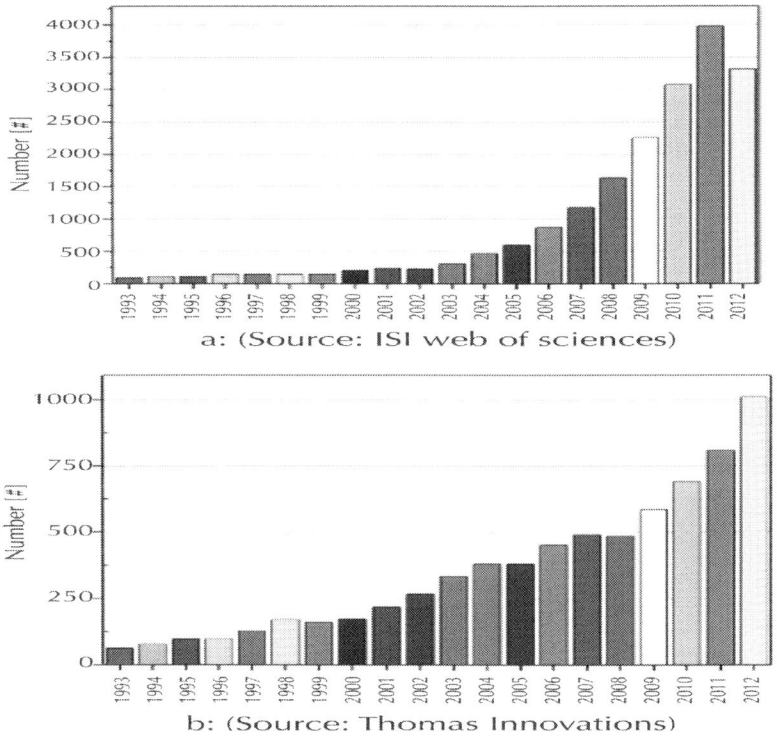

**Figure 1:** Citation trends of (a) publications and (b) patents on bio-based polymers in recent years.

Bio-based polymers offer important contributions by reducing the dependence on fossil fuels and through the related positive environmental impacts such as reduced carbon dioxide emissions. The legislative landscape is also changing where bio-based products are being favored through initiatives such as the *Lead Market Initiative* (European Union) and *BioPreferred* (USA). As a result, there is a worldwide demand for replacing petroleum-derived raw materials with renewable resource-based raw materials for the production of polymers.

The first generation of bio-based polymers focused on deriving polymers from agricultural feedstocks such as corn, potatoes, and other carbohydrate feedstocks. However, the focus has shifted in recent years due to a desire to move away from food-based resources and significant breakthroughs in biotechnology. Bio-based polymers similar to conventional polymers are produced by bacterial fermentation processes by synthesizing the building blocks (monomers) from renewable resources, including lignocellulosic biomass (starch and cellulose), fatty acids, and organic waste. Natural bio-based polymers are the other class of bio-based polymers which are found naturally, such as proteins, nucleic acids, and polysaccharides (collagen, chitosan, etc.). These bio-based polymers have shown enormous growth in recent years in terms of technological developments and their commercial applications. There are three principal ways to produce bio-based polymers using renewable resources:

1   Using natural bio-based polymers with partial modification to meet the requirements (e.g., starch)

2   Producing bio-based monomers by fermentation/conventional chemistry followed by polymerization (e.g., polylactic acid, polybutylene succinate, and polyethylene)

3   (3) Producing bio-based polymers directly by bacteria (e.g., polyhydroxyalkanoates).

In this paper, an overview of bio-based polymers made from renewable resources and natural polymers derived from plant and animal origins is presented. The review will focus on the preparation, properties, applications, and future trends for bio-based polymers. This paper discusses the use of renewable resources such as lignocellulosic biomass to create monomers and polymers that can replace petroleum-based polymers, such as polyester, polylactic acids, and other natural bio-based polymers, which are presented in Figure 2.

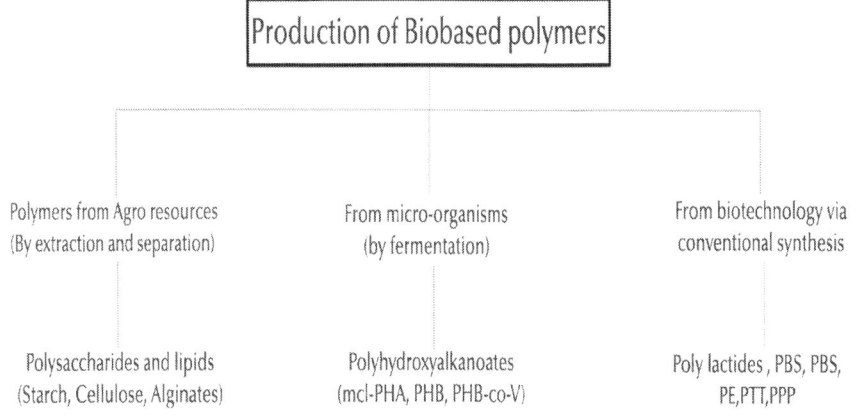

**Figure 2:** Most common categories of bio-based polymers produced by various processes. From Luc and Eric (2012).

# Polylactic Acid

Polylactic acid (PLA) has been known since 1845 but not commercialized until early 1990. PLA belongs to the family of aliphatic polyesters with the basic constitutional unit lactic acid. The monomer lactic acid is the hydroxyl carboxylic acid which can be obtained via bacterial fermentation from corn (starch) or sugars obtained from renewable resources. Although other renewable resources can be used, corn has the advantage of providing a high-quality feedstock for fermentation which results in a high-purity lactic acid, which is required for an efficient synthetic process. l-lactic acid or d-lactic acid is obtained depending on the microbial strain used during the fermentation process.

PLA can be synthesized from lactic acid by direct polycondensation reaction or ring-opening polymerization of lactide monomer. However, it is difficult to obtain high molecular weight PLA via polycondensation reaction because of water formation during the reaction. Nature Works LLC (previously Cargill Dow LLC) has developed a low-cost continuous process for the production of PLA (Erwin et al. 2007). In this process, low molecular weight pre-polymer lactide dimers are formed during a condensation process. In the second step, the pre-polymers are converted into high molecular weight PLA via ring-opening polymerization with selected catalysts. Depending on the ratio and stereochemical nature of the monomer (l or d), various types of PLA and PLA copolymers can be obtained. The final properties of PLA produced are highly dependent on the ratio of the d and l forms of the lactic acid which are listed in Table 1 for various blend ratios (Garlotta 2001).

**Table 1:** Variation in glass transition and melting temperature of PLA with various ratios of L-monomer composition

| Copolymer ratio | Glass transition ($T_g$),°C | Melting temperature ($T_m$),°C |
|---|---|---|
| 100:0 (l/dl)-PLA | 63 | 178 |
| 95:5 (l/dl)-PLA | 59 | 164 |
| 90:10 (l/dl)-PLA | 56 | 150 |
| 85:15 (l/dl)-PLA | 56 | 140 |
| 80:20 (l/dl)-PLA | 56 | 125 |

Babu *et al.*

Babu *et al. Progress in Biomaterials* 2013 2:8, doi:10.1186/2194-0517-2-8

PLA is a commercially interesting polymer as it shares some similarities with hydrocarbon polymers such as polyethylene terephthalate (PET). It has many unique characteristics, including good transparency, glossy appearance, high rigidity, and ability to tolerate various types of processing conditions.

PLA is a thermoplastic polymer which has the potential to replace traditional polymers such as PET, PS, and PC for packaging to electronic and automotive applications (Majid et al. 2010). While PLA has similar mechanical properties to traditional polymers, the thermal properties are not attractive due to low $T_g$ of 60°C. This problem can be overcome by changing the stereochemistry of the polymer and blending with other polymers and processing aids to improve the mechanical properties, e.g., varying the ratio of l and d isomer ratio strongly influences the crystallinity of the final polymer. However, much more work is required to improve the properties of PLA to suit various applications.

Currently, Nature Works LLC, USA, is the major supplier of PLA sold under the brand name Ingeo, with a production capacity of 100,000 ton/year. There are other manufactures of PLA based in the USA, Europe, China, and Japan developing various grades of PLA suitable for different industrial sectors such as automobile, electronics, medical devices, and commodity applications, which are mentioned in Table 2) (Doug 2010; Ravenstijn 2010).

**Table 2:** Global suppliers of PLA

| Company | Location | Brand name | Production/ planned capacity |
|---------|----------|------------|------------------------------|
| | | | **(kton/year)** |
| Nature Works | USA | Ingeo | 140 (by 2013) |
| Futerro | Belgium | Futerro | 1.5 (by 2010) |
| Tate & Lyle | Netherlands | Hycail | 0.2 (by 2012) |
| Purac | Netherlands | Purasorb | 0.05 |
| Hiusan Biosciences | China | Hisun | 5 |
| Jiangsu Jiulding | China | | 5 |
| Teijin | Japan | Biofront | 1 |
| Toyobo | Japan | Vylocol | 0.2 |
| Synbra | Netherlands | Biofoam | 50 |

Babu *et al.*

Babu *et al. Progress in Biomaterials* 2013 2:8, doi:10.1186/2194-0517-2-8

PLA is widely used in many day-to-day applications. It has been mainly used in food packing (including food trays, tableware such as plates and cutlery, water bottles, candy wraps, cups, etc.). Although PLA has one of the highest heat resistances and mechanical strengths of all bio-based polymers, it is still not suitable for use in electronic devices and other engineering applications. NEC Corporation (Japan) recently produced a PLA with carbon and kenaf fibers with improved thermal and flame retardancy properties. Fujitsu (Japan) developed a polycarbonate blend with PLA to make computer housings. In recent years, PLA has been employed as a membrane material for use in automotive and chemical industry.

The ease of melt processing has led to the production of PLA fibers, which are increasingly accepted in a wide variety of textiles from dresses to sportswear, furnishing to drapes, and soft nonwoven baby wipes to tough landscape textiles. These textiles can outperform traditional textiles made from synthetic counterparts. Bioresorbable scaffolds produced with PLA and various PLA blends are used in implants for growing living cells. The US Food and Drug Administration (FDA) has approved the use of PLA for certain human clinical applications (Dorozhkin 2009; Garlotta 2001). In addition, PLA-based materials have been used for bone support splints. Applications of PLA-based polymers in various fields are listed in Table 3.

**Table 3:** Application of PLA and their blends in various fields

| Polymer | Applications | Reference |
|---|---|---|
| PLGA/PGA | Ovine pulmonary valve replacement | Williams et al. 1999; Sodian et al. 1999, 2000; Cheng et al.2009 |
| PLA/chitosan PLA/PLGA/ chitosan PLA | Drug carrier/drug release | Jeevitha and Kanchana 2013; Jayanth and Vinod 2012; Nagarwal et al. 2010; Chandy et al. 2000; Valantin et al. 2003 |
| PLGA and copolymers | Degradable sutures | Rajev 2000 |
| PLA/HA composites | Porous scaffolds for cellular applications | Jung-Ju et al. 2012 |
| PLA-CaP and PLGA-CaP | Bone fixation devices, plates, pins, screws, and wires, orthopedic applications | Huan et al. 2012 |
| PDLLA | Coatings on metal implants | Schmidmaier et al. 2001 |

| PLA/PLGA | Use in cell-based gene therapy for cardiovascular diseases, muscle tissues, bone and cartilage regeneration, and other treatments of cardiovascular and neurological conditions | Coutu et al. 2009; Kellomaki et al. 2000; Papenburg et al.2009 |
|---|---|---|
| PLA and PLA blends | Packaging films, commodity containers, electrical appliances, mobile phone housings, floor mats, automotive spare parts | Rafael et al. 2010 |
| PLA | Textile applications | Gupta et al. 2007; Avinc and Akbar 2009 |

PLGA, polylactic acid-co-glycolic acid; CaP, calcium phosphates; HA, hydroxyapatite.

Babu *et al.*

Babu *et al. Progress in Biomaterials* 2013 2:8, doi:10.1186/2194-0517-2-8

# Polyhydroxyalkanoates

Polyhydroxyalkanoates (PHAs) are a family of polyesters produced by bacterial fermentation with the potential to replace conventional hydrocarbon-based polymers. PHAs occur naturally in a variety of organisms, but microorganisms can be employed to tailor their production in cells. Polyhydroxybutyrate (PHB), the simplest PHA, was discovered in 1926 by Maurice Lemoigne as a constituent of the bacterium *Bacillus megaterium* (Lemoigne 1923).

PHA can be produced by varieties of bacteria using several renewable waste feedstocks. A generic process to produce PHA by bacterial fermentation involves fermentation, isolation, and purification from fermentation broth. A large fermentation vessel is filled with mineral medium and inoculated with a seed culture that contains bacteria. The feedstocks include cellulosics, vegetable oils, organic waste, municipal solid waste, and fatty acids depending on the specific PHA required. The carbon source is fed into the vessel until it is consumed and cell growth and PHA accumulation is complete. In general, a minimum of 48 h is required for fermentation time. To isolate and purify PHA, cells are concentrated, dried, and extracted with solvents such as acetone or chloroform. The residual cell debris is removed from the solvent containing dissolved PHA by solid-liquid separation process. The PHA is then precipitated by the addition of an alcohol (e.g., methanol) and recovered by a precipitation process (Kathiraser et al. 2007).

More than 150 PHA monomers have been identified as the constituents of PHAs (Steinbüchel and Valentin 1995). Such diversity allows the production of bio-based polymers with a wide range of properties, tailored for specific applications. Poly-3-hydroxybutyrate was the first bacterial PHA identified. It has received the greatest attention in terms of pathway characterization and industrial-scale production. It possesses similar thermal and mechanical properties to those of polystyrene and polypropylene (Savenkova et al. 2000). However, due to its slow crystallization, narrow processing temperature range, and tendency to 'creep', it is not attractive for many applications, requiring development in order to overcome these shortcomings (Reis et al. 2008). Several companies have developed PHA copolymers with typically 80% to 95% (R)-3-hydroxybutyric acid monomer and 5% to 20% of a second monomer in order to improve the properties of PHAs. Some specific examples of PHAs include the following:

1.  Poly(3HB): Poly(3-hydroxybutyrate)

2.  Poly(3HB-co-3HV):                Poly(3-hydroxybutyrate-co-3-
    hydroxyvalerate), PHBV

3.  Poly(3-HB-co-4HB):               Poly(3-hydroxybutyrate-co-4-
    hydroxybutyrate)

4.  Poly(3HB-co-3HH):                Poly(3-hydroxyoctanoate-co-
    hydroxyhexanoate)

5.  Poly(3HO-co-3HH):                Poly(3-hydroxyoctanoate-co-
    hydroxyhexanoate)

6.  Poly (4-HB):                     Poly(4-hydroxybutyrate).

The copolymer poly(3HB-co-3HV) has a much lower crystallinity, decreased stiffness and brittleness, and increased tensile strength and toughness compared to poly(3HB) while remaining biodegradable.

It also has a higher melt viscosity, which is a desirable property for extrusion and blow molding (Hanggi 1995).

The first commercial plant for PHBV was built in the USA in a joint venture between Metabolix and Archer Daniels Midland. However, the joint venture between these two companies ended in 2012. Currently, Tianan Biologic Material Co. in China is the largest producer of PHB and PHB copolymers. Tianan's PHBV contains about 5% valerate which improves the flexibility of the polymer. Tainjin Green Biosciences, China, invested along with DSM to build a production plant with 10-kton/year capacity to produce PHAs for packing and biomedical applications (DSM press release 2008). The current global manufacturers of PHB-based polymers are listed in Table 4 (Doug 2010; Ravenstijn 2010).

**Table 4:** Global suppliers of various types of PHAs

| Company | Location | Brand name | Production/ planned capacity (kton/ year) |
|---|---|---|---|
| Bio-on | Italy | Minerv | 10 |
| Kaneka | Singapore | | 10 (by 2013) |
| Meredian | USA | | 13.5 |
| Metabolix | USA | Mirel | 50 |
| Mitsubishi Gas Chemicals | Japan | Biogreen | 0.05 |
| PHB Industrial S/A | Brazil | Biocycle | 0.05 |
| Shenzen O'Bioer | China | | |
| TEPHA | USA | ThephaFLEX/ ThephELAST | |
| Tianan Biological Materials | China | Enmat | 2 |
| Tianjin Green Biosciences | China | Green Bio | 10 |

| Tianjin Northern Food | China | | |
|---|---|---|---|
| Yikeman Shandong | China | | 3 |

Babu *et al.*

Babu *et al. Progress in Biomaterials* 2013 2:8, doi:10.1186/2194-0517-2-8

PHA polymers are thermoplastic, and their thermal and mechanical properties depend on their composition. The $Tg$ of the polymers varies from −40°C to 5°C, and the melting temperatures range from 50°C to 180°C, depending on their chemical composition (McChalicher and Srienc 2007). PHB is similar in its material properties to polypropylene, with a good resistance to moisture and aroma barrier properties. Polyhydroxybutyric acid synthesized from pure PHB is relatively brittle and stiff. PHB copolymers, which may include other fatty acids such as beta-hydroxyvaleric acid, may be elastic (McChalicher and Srienc 2007).

PHAs can be processed in existing polymer-processing equipment and can be converted into injection-molded components: film and sheet, fibers, laminates, and coated articles; nonwoven fabrics, synthetic paper products, disposable items, feminine hygiene products, adhesives, waxes, paints, binders, and foams. Metabolix has received FDA clearance for use of PHAs in food contact applications. These materials are suitable for a wide range of food packing applications including caps and closures, disposable items such as forks, spoons, knives, tubs, trays, and hot cup lids, and products such as housewares, cosmetics, and medical packaging (Philip et al. 2007).

PHA and its copolymers are widely used as biomedical implant materials. Various applications of PHA and their polymer blends are listed in Table 5. These include sutures, suture fasteners, meniscus repair devices, rivets, bone plates, surgical mesh, repair patches, cardiovascular patches, tissue repair patches, and stem cell growth. Changing the PHA composition allows the manufacturer to tune the properties such as biocompatibility and polymer degradation time within desirable time frames under specific conditions. PHAs can also be used in drug delivery due to their biocompatibility and controlled degradability. Only a few examples of PHAs have been evaluated for this type of applications, and it remains an important area for exploitation (Tang et al. 2008).

**Table 5:** Application of PHAs and their blends in various fields

| PHA polymer type | Applications | Reference |
|---|---|---|
| P(3HB), P(3HB-co-3HHX) and blends | Scaffolds, nerve regeneration, soft tissue, artificial esophagus, drug delivery, skin regeneration, food additive | Yang et al. 2002; Chen and Qiong 2005; Bayram and Denbas2008; Tang et al. 2008; Clarinval and Halleux 2005 |
| mcl-PHA/scl-PHA | Cardiac tissue engineering, drug delivery, cosmetics, drug molecules | Sodian et al. 2000; Wang et al. 2003; de Roo et al. 2002; Zhao et al. 2003; Ruth et al. 2007 |
| P(4HB) and P(3HO) | Heart valve scaffolds, food additive | Clarinval and Halleux 2005; Valappil et al. 2006 |
| P(3HB-co-4HB), P(3HB-co-3HV) | Drug delivery, scaffolds, artificial heart values, patches to repair gastrointestinal tracts, sutures | Türesin et al. 2001; Williams et al. 1999; Chen et al. 2008; Freier et al. 2002; Kunze et al. 2006; Volova et al. 2003 |
| PHB, Mirel P103 | Commodity applications, shampoo and cosmetic bottles, cups and food containers | Philip et al. 2007; Amass et al. 1998; Walle et al. 2001 |

Babu et al.

Babu *et al. Progress in Biomaterials* 2013 2:8, doi:10.1186/2194-0517-2-8

# Polybutylene Succinate

Polybutylene succinate (PBS) is an aliphatic polyester with similar properties to those of PET. PBS is produced by condensation of succinic acid and 1,4-butanediol. PBS can be produced by either monomers derived from petroleum-based systems or the bacterial fermentation route. There are several processes for producing succinic acid from fossil fuels. Among them, electrochemical synthesis is a common process with high yield and low cost. However, the fermentation production of succinic acid has numerous advantages compared to the chemical process. Fermentation process uses renewable resources and consumes less energy compared to chemical process. Several companies (solely or in partnership) are now scaling bio-succinate production processes which have traditionally suffered from poor productivity and high downstream processing costs. Mitsubishi Chemical (Japan) has developed biomass-derived succinic acid in collaboration with Ajinomoto to commercialize bio-based PBS. DSM and Roquette are developing a commercially feasible fermentation process for the production of succinic acid 1,4-butanediol and subsequent production of PBS. Myriant and Bioamber have developed a fermentation technology to produce monomers. There are several companies around the world developing technologies for the production of PBS, as listed in Table 6, including North America and China (Doug 2010; Ravenstijn 2010).

**Table 6:** Global producers of PBS

| Company | Location | Brand name/ polymer type | Production/ planned capacity (kton/year) |
|---|---|---|---|
| BASF | Germany | PBS | |
| Dupont de Nemours | USA | PBST | |
| Hexing Chemical | China | PBS | 3 |
| Ube | Japan | NA | NA |
| IPC-CAS | China | PBS, PBSA | 5 |

| IRE Chemical | Korea | Enpol, PBS, PBSA | 3.5 |
|---|---|---|---|
| Kingfa | China | PBSA | 1 |
| Mitsubishi Gas Chemical | Japan | PBS, PES, PBSLa | 3 |
| Showa | Japan | Bionelle PBS, PBSA, PBS | 3 |
| SK Chemicals | Korea | Skygreen | NA |
| DSM | Netherlands | NA | NA |

NA, not available; PBSA, poly(butylene succinate adipate).

Babu *et al.*

Babu *et al. Progress in Biomaterials* 2013 2:8, doi:10.1186/2194-0517-2-8

Conventional processes for the production of 1,4-butanediol use fossil fuel feedstocks such as acetylene and formaldehyde. The bio-based process involves the use of glucose from renewable resources to produce succinic acid followed by a chemical reduction to produce butanediol. PBS is produced by transesterification, direct polymerization, and condensation polymerization reactions. PBS copolymers can be produced by adding a third monomer such as sebacic acid, adipic acid, and succinic acid which is also produced by renewable resources (Bechthold et al. 2008).

PBS is a semicrystalline polyester with a melting point higher than that of PLA. Its mechanical and thermal properties depend on the crystal structure and the degree of crystallinity (Nicolas et al.2011). PBS displays similar crystallization behavior and mechanical properties to those of polyolefin such as polyethylene. It has a good tensile and impact strength with moderate rigidity and hardness. The $Tg$ is approximately $-32°C$, and the melting temperature is approximately 115°C. In comparison with PLA, PBS is tougher in nature but with a lower rigidity and Young›s modulus. By changing the monomer composition, mechanical properties can be tuned to suit the required application (Liu et al. 2009a, b).

PBS and their blends have found commercial applications in agriculture, fishery, forestry, construction, and other industrial fields which are listed in Table 7. For example, PBS has been employed as mulch film, packaging, and flushable hygiene products and also used as a non-migrant plasticizer for polyvinyl chloride (PVC). In addition, it is used in foaming and food packaging application. The relatively poor mechanical flexibility of PBS limits the applications of 100% PBS-based products. However, this can be overcome by blending PBS with PLA or starch to improve the mechanical properties significantly, providing properties similar to that of polyolefin (Eslmai and Kamal 2013; Zhao et al. 2010).

**Table 7:** Applications of PBS and their blends

| Polymer type | Applications | Reference |
|---|---|---|
| PBS/PLA blend | Packaging films, dishware, fibers, medical materials | Weraporn et al. 2011; Liu et al. 2009 a, b; Bhatia et al. 2007; Lee and Wang 2006 |
| PBS and blends | Drug encapsulation systems | Cornelia et al. 2011 |
| PBS/starch | Barrier films | Jian-Bing et al. 2011 |
| PBS and copolymers | Industrial applications | Jun and Bao-Hua 2010 a, b |
| PBS ionomers | Orthopedic applications | Jung et al. 2009 |

Babu *et al.*

Babu *et al. Progress in Biomaterials* 2013 **2**:8, doi:10.1186/2194-0517-2-8

# Bio-Polyethylene

Polyethylene (PE) is an important engineering polymer traditionally produced from fossil resources. PE is produced by polymerization of ethylene under pressure, temperature, in the presence of a catalyst. Traditionally, ethylene is produced through steam cracking of naphtha or heavy oils or ethanol dehydration. With increases in oil prices, microbial PE or green PE is now being manufactured from dehydration of ethanol produced by microbial fermentation. The concept of producing PE from bioethanol is not a particularly new one. In the 1980s, Braskem made bio-PE and bio-PVC from bioethanol. However, low oil prices and the limitations of the biotechnology processes made the technology unattractive at that time (de Guzman 2010).

Currently, bio-PE produced on an industrial scale from bioethanol is derived from sugarcane. Bioethanol is also derived from biorenewable feedstocks, including sugar beet, starch crops such as maize, wood, wheat, corn, and other plant wastes through microbial strain and biological fermentation process. In a typical process, extracted sugarcane juice with high sucrose content is anaerobically fermented to produce ethanol. At the end of the fermentation process, ethanol is distilled in order to remove water and to yield azeotropic mixture of hydrous ethanol. Ethanol is then dehydrated at high temperatures over a solid catalyst to produce ethylene and, subsequently, polyethylene (Guangwen et al. 2007; Luiz et al. 2010).

Bio-based polyethylene has exactly the same chemical, physical, and mechanical properties as petrochemical polyethylene. Braskem (Brazil) is the largest producer of bio-PE with 52% market share, and this is the first certified bio-PE in the world. Similarly, Braskem is developing other bio-based polymers such as bio-polyvinyl chloride, bio-polypropylene, and their copolymers with similar industrial technologies. The current Braskem bio-based PE grades are mainly targeted towards food packing, cosmetics, personal care, automotive parts, and toys. Dow Chemical (USA) in cooperation with Crystalsev is the second largest producer of bio-PE with 12% market share. Solvay (Belgium), another producer of bio-PE, has 10% share in the current market. However, Solvay is a leader in the production of bio-PVC with similar industrial technologies. China Petrochemical Corporation also plans to set up production facilities in China to produce bio-PE from bioethanol (Haung et al. 2008).

Bio-PE can replace all the applications of current fossil-based PE. It is widely used in engineering, agriculture, packaging, and many day-to-day commodity applications because of its low price and good performance. Table 8 shows applications of bio-PE in different fields where it can replace conventional PE.

**Table 8:** Application of bio-PE polymer and their blends

| Polymer type | Applications | Reference |
|---|---|---|
| Bio-PE | Plastics bags, milk and water bottles, food packaging films, toys | Vona et al. 1965; Aamer et al. 2008 |
| Bio-PE and blends | Agricultural mulch films | Kasirajan and Ngouajio 2012 |

Babu *et al.*

Babu *et al. Progress in Biomaterials* 2013 2:8, doi:10.1186/2194-0517-2-8

# Bio-Based Natural Polymers

This group consists of naturally occurring polymers such as cellulose, starch, chitin, and various polysaccharides and proteins. These materials and their derivatives offer a wide range of properties and applications. In this section, some of the natural bio-based polymers and their applications in various fields are discussed.

# Starch

Starch is a unique bio-based polymer because it occurs in nature as discrete granules. Starch is the end product of photosynthesis in plants - a natural carbohydrate-based polymer that is abundantly available in nature from various sources including wheat, rice, corn, and potato. Essentially, starch consists of the linear polysaccharide amylose and the highly branched polysaccharide amylopectin. In particular, thermoplastic starch is of growing interest within the industry. The thermal and mechanical properties of starch can vary greatly and depend upon such factors as the amount of plasticizer present. The $Tg$ varies between $-50°C$ and $110°C$, and the modulus is similar to polyolefins (Jane1995). Several challenges exist in producing commercially viable starch plastics. Starch›s molecular structure is complex and partly nonlinear, leading to issues with ductility. Starch and starch thermoplastics suffer from the phenomenon of retrogradation - a natural increase in crystallinity over time, leading to increased brittleness. Plasticizers need to be found to create starch plastics with mechanical properties comparable to polyolefin-derived packaging. Plasticized starch blends and composites and/or chemical modifications may overcome these issues, creating biodegradable polymers with sufficient mechanical strength, flexibility, and water barrier properties for commercial packaging and consumer products (Maurizio et al. 2005).

Novamont is one of the leading companies in processing starch-based products (Li et al. 2009). The company produces various types of starch-based products using proprietary blend formulations. There are other companies around the world producing starch-based products in a similar scale for various applications, which are listed in Table 9 (Doug 2010; Ravenstijn 2010).

**Table 9:** Global suppliers of starch-based products

| Company | Location | **Brand name** | **Production/ planned capacity (kton/year)** |
|---|---|---|---|
| Novamont | Italy | Mater-Bi | 120 |
| Japan Corn Starch | Japan | Ever Corn | NA |
| Biotec | Germany | Bioplast | NA |
| Rodenberg | Netherlands | Solanyl | 50 |
| BIOP | Germany | Biopar | 5 |
| Plantic | Australia | Plantic | 7.5 |
| Wuhan Huali Environment Protection Sci. & Tech | China | PSM | 15 |
| Biograde | China | Cardia | 3 |
| PSM | USA | Plaststarch | NA |
| Livan | Canada | Livan | 10 |

Babu et al.

Babu *et al. Progress in Biomaterials* 2013 2:8, doi:10.1186/2194-0517-2-8

Applications of thermoplastic starch polymers include films, such as for shopping, bread, and fishing bait bags, overwraps, flushable sanitary product, packing materials, and special mulch films. Potential future applications could include foam loose-fill packaging and injection-molded products such as 'take-away' food containers. Starch and modified starches have a broad range of applications both in the food and non-food sectors. In Europe in 2002, the total consumption of starch and starch derivatives was approximately 7.9 million tons, of which 54% was used for food applications and 46% in non-food applications (Frost & Sullivan report 2009).

The largest users of starch in the European Union (30%) are the paper, cardboard, and corrugating industries (Frost & Sullivan report 2009). Other important fields of starch application are textiles, cosmetics, pharmaceuticals, construction, and paints, which are listed in Table 10. In the medium and long term, starch will play an increasing role in the field of 'renewable raw materials' for the production of biodegradable plastics, packaging material, and molded products.

**Table 10:** Application of starch and their blends in various fields

| Polymer type | Applications | Reference |
|---|---|---|
| Starch | Orthopedic implant devices as bone fillers | Ashammakhi and Rokkanen 1997 |
| Starch/ethylene vinyl alcohol/HA starch/ polycaprolactone blends | Bone replacement/ fixation implants, orthopedic applications | Mainil et al. 1997; Mendes et al. 2001; Marques and Reis2005 |
| Starch/cellulose acetate blends with methylmethacrylate and acrylic acid | Bone cements | Espigares et al. 2002 |
| Modified starch | Food applications | Jaspreet et al. 2007; Fuentes et al. 2010 |
| Starch derivatives | Drug delivery | Asha and Martins 2012 |

| Thermoplastic starch | Packaging, containers, mulch films, textile sizing agents, adhesives | Zhao et al. 2008; Maurizio et al. 2005; Ozdemir and Floros2004; Dave et al. 1999; Guo et al. 2005; Kumbar et al.2001; Li et al. 2011 |

Babu *et al.*

Babu *et al. Progress in Biomaterials* 2013 **2**:8, doi:10.1186/2194-0517-2-8

# Cellulose

Cellulose is the predominant constituent in cell walls of all plants. Cellulose is a complex polysaccharide with crystalline morphology. Cellulose differs from starch where glucose units are linked by β-1,4-glycosidic bonds, whereas the bonds in starch are predominantly α-1,4 linkages. The most important raw material sources for the production of cellulosic plastics are cotton fibers and wood. Plant fiber is dissolved in alkali and carbon disulfide to create viscose, which is then reconverted to cellulose in cellophane form following a sulfuric acid and sodium sulfate bath. There are currently two processes used to separate cellulose from the other wood constituents (Yan et al.2009). These methods, sulfite and pre-hydrolysis kraft pulping, use high pressure and chemicals to separate cellulose from lignin and hemicellulose, attaining greater than 97% cellulose purity. The main derivatives of cellulose for industrial purposes are cellulose acetate, cellulose esters (molding, extrusion, and films), and regenerated cellulose for fibers.

Cellulose is a hard polymer and has a high tensile strength of 62 to 500 MPa and elongation of 4% (Bisanda and Ansell 1992; Eichhorn et al. 2001). In order to overcome the inherent processing problems of cellulose, it is necessary to modify, plasticize, and blend with other polymers. The mechanical and thermal properties vary from blend to blend depending on the composition. The $T_g$ of cellulosic derivatives ranged between 53°C and 180°C (Picker and Hoag 2002).

Eastman Chemical is a major producer of cellulosic polymers. FKuR launched a biopolymer business in the year 2000 and has a capacity of 2,800 metric ton/year of various cellulosic compounds for different applications (Doug 2010). The major producers of cellulose-based compounds are listed in Table 11 (Doug 2010; Ravenstijn 2010).

**Table 11:** Global suppliers of cellulosic products

| Company | Location | Brand name |
|---|---|---|
| Innovia films | UK | Nature Flex |
| Eastman Chemical | USA | Tenite |
| FKuR | Germany | Biograde |
| Sateri | China | Sateri |

Babu *et al.*

Babu *et al. Progress in Biomaterials* 2013 **2**:8, doi:10.1186/2194-0517-2-8

There are three main groups of cellulosic polymers that are produced by chemical modification of cellulose for various applications. Cellulose esters, namely cellulose nitrate and cellulose acetate, are mainly developed for film and fiber applications. Cellulose ethers, such as carboxymethyl cellulose and hydroxyethyl cellulose, are widely used in construction, food, personal care, pharmaceuticals, paint, and other pharmaceutical applications (Kamel et al. 2008). Finally, regenerated cellulose is the largest bio-based polymer produced globally for fiber and film applications. Regenerated cellulose fibers are used in textiles, hygienic disposables, and home furnishing fabrics because of its thermal stability and modulus (Kevin et al. 2001).

Chemically pure cellulose can be produced using a certain type of bacteria. Bacterial cellulose is characterized by its purity and high strength. It can be used to produce articles with relatively high strength. Currently, applications for bacterial cellulose outside food and biomedical fields are rather limited because of its high price. The other applications include acoustic diaphragms, mining, paints, oil gas recovery, and adhesives. However, the low yields and high costs of bacterial cellulose represent barriers to large-scale industrial applications (Prashant et al. 2009). Table 12 summarizes the applications of cellulose and their compounds in different fields.

**Table 12:** Application of cellulose and their compounds in various fields

| Polymer type | Applications | Reference |
|---|---|---|
| Cellulose esters | Membranes for separation | Kumano and Fujiwara 2008 |
| Carboxylated methyl cellulose | Drug formulations, as binder for drugs, film-coating agent for drugs, ointment base | Chambin et al. 2004; Obae and Imada 1999; Westermark et al.1999; Hirosawa et al. 2000 |
| Cellulose acetate fibers | Wound dressings | Orawan et al. 2008; Abdelrahman and Newton 2011 |
| Hydroxyethyl cellulose | Spray for clothes polluted with pollen | Hori et al. 2005 |
| Modified celluloses, cellulose whiskers, microfibrous cellulose | Barrier films, water preservation in food packing | Amit and Ragauskas 2009 |
| Cellulose nanofibers | Textile applications | Zeeshan et al. 2013 |
| Cellulose particles | Chromatographic applications, chiral separations | Levison 1993; Arshady 1991a, b |

Babu *et al.*

Babu *et al. Progress in Biomaterials* 2013 **2**:8, doi:10.1186/2194-0517-2-8

# Chitin and Chitosan

Chitin and chitosan are the most abundant natural amino polysaccharide and valuable bio-based natural polymers derived from shells of prawns and crabs. Currently, chitin and chitosan are produced commercially by chemical extraction process from crab, shrimp, and prawn wastes (Roberts1997). The chemical extraction of chitin is quite an aggressive process based on demineralization by acid and deproteination by the action of alkali followed by deacetylated into chitosan (Roberts

1997). Chitin can also be produced by using enzyme hydrolysis or fermentation process, but these processes are not economically feasible on an industrial scale (Win and Stevens 2001). Currently, there are few industrial-scale plants of chitin and chitosan worldwide located in the USA, Canada, Scandinavia, and Asia (Ravi Kumar 2000).

Chitosan displays interesting characteristics including biodegradability, biocompatibility, chemical inertness, high mechanical strength, good film-forming properties, and low cost (Marguerite 2006; Virginia et al. 2011; Liu et al. 2012). Chitosan is being used in a vast array of widely varying products and applications ranging from pharmaceutical and cosmetic products to water treatment and plant protection. For each application, different properties of chitosan are required, which changes with the degree of acetylation and molecular weight. Chitosan is compatible with many biologically active components incorporated in cosmetic product composition (Ravi Kumar 2000). Due to its low toxicity, biocompatibility, and bioactivity, chitosan has become a very attractive material in such diverse applications as biomaterials in medical devices and as a pharmaceutical ingredient (Bae and Moo-Moo 2010; Ramya et al. 2012). Chitosan has application in shampoos, rinses, and permanent hair-coloring agents. Chitosan and its derivatives also have applications in the skin care industry. Chitosan can function as a moisturizer for the skin, and because of its lower costs, it might compete with hyaluronic acid in this application (Bansal et al. 2011; Valerie and Vinod 1998; Hafdani and Sadeghinia 2011).

# Pullulan

Pullulan is a linear water-soluble polysaccharide mainly consisting of maltotriose units connected by α-1,6 glycosidic units. Pullulan was first reported by Bauer (1938) and is obtained from the fermentation broth of *Aureobasidium pullulans*. Pullulan is produced by a simple fermentation process using a number of feedstocks containing simple sugars (Bernier 1958; Catley 1971; Sena et al.2006). Pullulan can be chemically modified to produce a polymer that is either less soluble or completely insoluble in water. The unique properties of this polysaccharide are due to its characteristic glycosidic linking. Pullulan is easily chemically modified to reduce the water solubility or to develop pH sensitivity, by introducing functional reactive groups, etc. Due to its high water solubility and low viscosity, pullulan has numerous commercial applications including use as a food additive, a flocculant, a blood plasma substitute, an adhesive, and a film (Zajic and LeDuy 1973; Singh et al. 2008; Cheng et al. 2011). Pullulan can be formed into molding articles which can resemble conventional polymers such as polystyrene in their transparency, strength, and toughness (Leathers 2003).

Pullulan is extensively used in the food industry. It is a slow-digesting macromolecule which is tasteless as well as odorless, hence its application as a low-calorie food additive providing bulk and texture. Pullulan possesses oxygen barrier property and good moisture retention, and also, it inhibits fungal growth. These properties make it an excellent material for food preservation, and it is used extensively in the food industry (Conca and Yang 1993). In recent years, pullulan has also been studied for biomedical applications in various aspects, including targeted drug and gene delivery, tissue engineering, wound healing, and even in diagnostic imaging medium (Rekha and Chrndra2007). Other emerging markets for pullulan include oral care products (Barkalow et al. 2002) and formulations of capsules for dietary supplements and pharmaceuticals (Leathers 2003), leading to increased demand for this unique biopolymer.

# Collagen and Gelatin

Collagen is the major insoluble fibrous protein in the extracellular matrix and in connective tissue. In fact, it is the single most abundant protein in the animal kingdom. There are at least 27 types of collagens, and the structures all serve the same purpose: to help tissues withstand stretching. The most abundant sources of collagen are pig skin, bovine hide, and pork and cattle bones. However, the industrial use of collagen is obtained from nonmammalian species (Gomez-Guille et al. 2011). Gelatin is obtained through the hydrolysis of collagen. The degree of conversion of collagen into gelatin depends on the pretreatment, function of temperature, pH, and extraction time (Johnston-Banks 1990).

Collagen is one of the most useful biomaterials due to its biocompatibility, biodegradability, and weak antigenicity (Maeda et al. 1999). The main application of collagen films in ophthalmology is as drug delivery systems for slow release of incorporated drugs (Rubin et al. 1973). It was also used for tissue engineering including skin replacement, bone substitutes, and artificial blood vessels and valves (Lee et al. 2001).

The classical food, photographic, cosmetic, and pharmaceutical applications of gelatin is based mainly on its gel-forming properties. Recently in the food industry, an increasing number of new applications have been found for gelatin in products in line with the growing trend to replace synthetic agents with more natural ones (Gomez-Guille et al. 2011). These include emulsifiers, foaming agents, colloid stabilizers, biodegradable film-forming materials, and microencapsulating agents.

# Alginates

Alginate is a linear polysaccharide that is abundant in nature as it is synthesized by brown seaweeds and by soil bacteria (Draget et al. 1997). Sodium alginate is the most commonly used alginate form in the industry since it is the first by-product of algal purification (Draget 2000). Sodium alginate consists of α-l-guluronic acid residues (G blocks) and β-d-mannuronic acid residues (M blocks), as well as segments of alternating guluronic and mannuronic acids.

Although alginates are a heterogeneous family of polymers with varying content of G and M blocks depending on the source of extraction, alginates with high G content have far more industrial importance (Siddhesh and Edgar 2012). The acid or alkali treatment processes used to make sodium alginate from brown seaweeds are relatively simple. The difficulties in processing arise mainly from the separation of sodium alginate from slimy residues (Black and Woodward 1954). It is estimated that the annual production of alginates is approximately 38,000 tons worldwide (Helgerud et al.2009).

Alginates have various industrial uses as viscosifiers, stabilizers, and gel-forming, film-forming, or water-binding agents (Helga and Svein 1998). These applications range from textile printing and manufacturing of ceramics to production of welding rods and water treatment (Teli and Chiplunkar1986; Qin et al. 2007; Xie et al. 2001). The polymer is soluble in cold water and forms thermostable gels. These properties are utilized in the food industry in products such as custard creams and restructured food. The polymer is also used as a stabilizer and thickener in a variety of beverages, ice creams, emulsions, and sauces (Iain et al. 2009).

Alginates are widely used as a gelling agent in pharmaceutical and food applications. Studies into their positive effects on human health have broadened recently with the recognition that they have a number of potentially beneficial physiological effects in the gastrointestinal tract (Peter et al. 2011; Mandel et al. 2000). Alginate-containing wound dressings are commonly used, especially in making hydrophilic gels over wounds which can produce comfortable, localized hydrophilic environments in healing wounds (Onsoyen 1996). Alginates are used in controlled drug delivery, where the rate of drug release depends on the type and molecular weight of alginates used (Alexnader et al. 2006; Goh et al. 2012). Additionally, dental impressions made with alginates are easy to handle for both dentist and patient as they fast set at room temperature and are cost-effective (Onsoyen 1996). Recent studies show that alginates can be effective in treating obesity, and currently, various functional alginates are being evaluated in human clinical trials (Georg et al. 2012).

## Current Status and Future Trends

The use of bio-based feedstocks in the chemical sector is not a novel concept. They have been industrially feasible on a large scale for more than a decade. However, the price of oil was so cost-effective, and the development of oil-based products created so many opportunities that bio-based products were not prioritized at the time. Several factors, such as the limitations and uncertainty in supplies of fossil fuels, environmental considerations, and technological developments, accelerated the advancement of bio-based polymers and products. It took more than a century to evolve the fossil fuel-based chemical industry; however, the bio-based polymer industry is already catching up with fossil fuel-based chemical industry, which has augmented in the last 20 years. Thanks to advancements in white biotechnology, the production of bio-based polymers and other chemicals from renewable resources has become a reality. The first-generation technologies mainly focused on food resources such as corn, starch, rice, etc. to produce bio-based polymers. As the food-versus-fuel debate ascended, the focus of technologies diverted to cellulose-based feedstocks, focusing on waste from wood and paper, food industries, and even stems and leaves and solid municipal waste streams. More and more of these technologies are already in the pipeline to align with the abovementioned waste streams; however, it may take another 20 years to develop the full spectrum of chemicals based on these technologies (Michael et al. 2011).

Challenges that need to be addressed in the coming years include management of raw materials, performance of bio-based materials, and their cost for production. Economy of scale will be one of the main challenges for production of bio-based monomers and bio-based polymers from renewable sources. Building large-scale plants can be difficult due to the lack of experience in new technologies and estimation of supply/demand balance. In order to make these technologies economically viable, it is very important to develop (1) logistics for biomass feedstocks, (2) new manufacturing routes by replacing existing methods with high yields, (3) new microbial strains/enzymes, and (4) efficient downstream processing methods for recovery of bio-based products.

The current bio-based industry focus is mainly on making bio-versions of existing monomers and polymers. Performance of these products is well known, and it is relatively easy to replace the existing product with similar performance of bio-versions. All the polymers mentioned above often display similar properties of current fossil-based polymers. Recently, many efforts are seen towards introducing new bio-based polymers with higher performance and value. For example, Nature Works LLC has introduced new grades of PLA with higher thermal and mechanical properties. New PLA-tri block copolymers have been reported to behave like thermoplastic elastomer. Many developments are currently underway to develop various polyamides, polyesters, polyhydroxyaloknates, etc. with a high differentiation in their final properties for use in automotive, electronics, and biomedical applications.

The disadvantage of some of the new bio-based polymers is that they cannot be processed in all current processing equipment. There is vast knowledge on additive-based chemistry developed for improving the performance and processing of fossil fuel-based polymers, and this knowledge can be used to develop new additive chemistry to improve the performance and properties of bio-based polymers (Ray and

Bousmina 2005). For bio-based polymers like PLA and PHA, additives are being developed to improve their performance, by blending with other polymers or making new copolymers. However, the additive market for bio-based polymers is still very small, which makes it difficult to justify major development efforts according to some key additive supplier companies.

The use of nanoparticles as additives to enhance polymer performance has long been established for petroleum-based polymers. Various nano-reinforcements currently being developed include carbon nanotubes, graphene, nanoclays, 2-D layered materials, and cellulose nanowhiskers. Combining these nanofillers with bio-based polymers could enhance a large number of physical properties, including barrier, flame resistance, thermal stability, solvent uptake, and rate of biodegradability, relative to unmodified polymer resin. These improvements are generally attained at low filler content, and this nano-reinforcement is a very attractive route to generate new functional biomaterials for various applications.

Even though new bio-based polymers are produced on an industrial scale, there are still several factors which need to be determined for the long-term viability of bio-based polymers. It is expected that there will be feedstock competition as global demand for food and energy increases over time. Currently, renewable feedstocks used for manufacturing bio-based monomers and polymers often compete with requirements for food-based products. The expansion of first-generation bio-based fuel production will place unsustainable demands on biomass resources and is as much a threat to the sustainability of biochemical and biopolymer production as it is to food production (Michael et al.2011). Indeed the European commission has altered its targets downwards for first-generation biofuels since October 2012, indicating its preference for non-food sources of sugar for biofuel production (EurActiv.com 2012). Several initiatives are underway to use cellulose-based feedstocks for the production of usable sugars for biofuels, biochemicals, and biopolymers (Jong et al. 2010).

# CONCLUSIONS

Bio-based polymers are closer to the reality of replacing conventional polymers than ever before. Nowadays, bio-based polymers are commonly found in many applications from commodity to hi-tech applications due to advancement in biotechnologies and public awareness. However, despite these advancements, there are still some drawbacks which prevent the wider commercialization of bio-based polymers in many applications. This is mainly due to performance and price when compared with their conventional counterparts, which remains a significant challenge for bio-based polymers.

# AUTHORS' CONTRIBUTIONS

RPB contributed in writing the whole manuscript. KOC contributed in providing the information on applications and policy information of bio-based polymers. SR contributed in providing the outline for the manuscript. All authors read and approved the final manuscript.

# ACKNOWLEDGMENTS

RPB would like to acknowledge the financial support from the Environmental Protection Agency, Ireland, under grant no. 2008-ET-LS-1-S2.

# REFERENCES

1.   Aamer AS, Fariha H, Abdul H, Safia A (2008) Biological degradation of plastics: a comprehensive review. Biotechnol Adv 26:246-265

2.   Abdelrahman T, Newton H (2011) Wound dressings: principles and practice. Surgery 29:491-495

3.   Alexnader DA, Kong HJ, Mooney DJ (2006) Alginate hydrogels as biomaterials. Macromolecular Biosciences 6:623-633

4.   Amass W, Amass A, Tighe B (1998) A review of biodegradable polymers: uses, current developments in the synthesis and characterization of biodegradable polyesters, blends of biodegradable polymers and recent advances in biodegradation studies. Polymer International 47:89-144

5.  Amit S, Ragauskas AJ (2009) Water transmission barrier properties of biodegradable films based on cellulosic whiskers and xylan. Carbohydr Polym 78(2):357-360

6.  Arshady R (1991) Beaded polymer supports and gels: 2. Physicochemical criteria and functionalization. J Chromatogr 586:199-219

7.  Arshady R (1991) Beaded polymer supports and gels: 1. Manufacturing techniques. J Chromatogr 586:181-197

8.  Asha R, Martins E (2012) Recent applications of starch derivatives in nanodrug delivery. Carbohydr Polym 87(2):987-994

9.  Ashammakhi N, Rokkanen P (1997) Absorbable polyglycolide devices in trauma and bone surgery. Biomaterials 18(1):3-9

10. Avinc A, Akbar K (2009) Overview of poly (lactic acid) fibres. Part I: production, properties, performance, environmental impact, and end-use applications of poly (lactic acid) fibres. Fiber Chemistry 41(6):391-401

11. Bae KP, Moo-Moo K (2010) Applications of chitin and its derivatives in biological medicine. Int J Mol Sci 11:5152-5164

12. Bansal V, Pramod KS, Nitin S, Omprakask P, Malviya R (2011) Applications of chitosan and chitosan derivatives for drug delivery. Adva Biol Res 5:28-37

13. Barkalow DG, Chapedelaine AH, Dzija MJ (2002) Improved pullulan free edible film compositions and methods of making same. PCT International Application WO 02/43657, US 01/43397, 21 Nov.

14. Bauer R (1938) Physiology of Dematium pullulans de Bary. Zentralbl Bacteriol Parasitenkd Infektionskr Hyg Abt2 98:133-167

15. Bayram C, Denbas EB (2008) Preparation and characterization of triamcinolone acetonide-loaded poly(3-hydroxybutyrate-co-3-hydroxyhexanoate) (PHBHx) microspheres. J Bioactive and Compatible Polymer 23:334-347

16. Bechthold I, Bretz K, Kabasci S, Kopitzky R, Springer A (2008) Succinic acid: a new platform chemical from biobased polymers from renewable resources. Chemical Engg Technol 31:647-654

17. Bernier B (1958) The production of polysaccharides by fungi active in the decomposition of wood and forest litter. Can J Microbiol 4:195-204

18. Bhatia A, Gupta RK, Bhattacharaya SN, Choi HJ (2007) Compatibility of biodegradable PLA and PBS blends for packaging applications. Korea Aust Rheol J 19:125-131

19. Bisanda ETN, Ansell MP (1992) Properties of sisal-CNSL composites. J Mater Sci 27:1690-1700

20. Black WAP, Woodward FN (1954) Alginates from common British brown marine algae. In Natural plant hydrocolloids. Adv Chem Ser Am Chem Soc 11:83-91

21. Catley BJ (1971) Utilization of carbon sources by Pullularia pullulans for the elaboration of extracellular polysaccharides. Appl Microbiol 22:641-649

22. Chambin DC, Debray C, Rochat-Gonthier MH, Le MM, Pourcelot M (2004) Effects of different cellulose derivatives on drug release mechanism studied at a pre-formulation stage. J Controll Release 95(1):101-108

23. Chandy T, Das GS, Rao GH (2000) 5-Fluorouracil-loaded chitosan coated polylactic acid microspheres as biodegradable drug carriers for cerebral tumours. J Microencapsul 5:625-631

24. Chen GQ, Qiong W (2005) The application of polyhydroxyalkanoates as tissue engineering materials. Biomaterials 26:6565-6578

25. Chen GQ, Martin KP (2012) Plastics derived from biological sources: present and future: a technical and environmental review. Chem Rev 112:2082-2099

26. Chen QZ, Harding SE, Ali NN, Lyon AR, Boccaccini AR (2008) Biomaterials in cardiac tissue engineering: ten years of research survey. Materials Sci Eng: Reports 59:1-37

27. Cheng KC, Demirci A, Catchmark JM (2011) Pullulan: biosynthesis, production, and applications. Appl Microbiol Biotechnol 92:29-44

28. Cheng Y, Deng S, Chen P, Ruan R (2009) Polylactic acid (PLA) synthesis and modifications: a review. Front Chem China 4:259-264

29.  Clarinval AM, Halleux J (2005) Classification of biodegradable polymers. In: Smith R (ed) Biodegradable polymers for industrial applications, Woodhead, Cambridge.

30.  Conca KR, Yang TCS (1993) Edible food barrier coatings. In: Ching C, Kaplan DL, Thomas EL (eds) Biodegradable polymers and packaging, Technomic, Lancaster. pp 357-369

31.  Cornelia TB, Erkan TB, Elisabete DP, Rui LR, Nuno MN (2011) Performance of biodegradable microcapsules of poly(butylene succinate), poly(butylene succinate-co-adipate) and poly(butylene terephthalate-co-adipate) as drug encapsulation systems. Colloids Surf B Biointerfaces 84:498-507

32.  Coutu DL, Yousefi AM, Galipeau J (2009) Three-dimensional porous scaffolds at the crossroads of tissue engineering and cell-based gene therapy. J Cell Biochem 108:537-546

33.  Dave AM, Mehta MH, Aminabhavi TM, Kulkarni AR, Soppimath KS (1999) A review on controlled release of nitrogen fertilizers through polymeric membrane devices. Polymer- Plastics Technol Eng 38:675-711

34.  de Roo G, Kellerhals MB, Ren Q, Witholt B, Kessler B (2002) Production of chiral R-3-hydroxyalkanoic acids and R-3-hydroxyalkanoic acid methylesters via hydrolytic degradation of polyhydroxyalkanoate synthesized by pseudomonads. Biotechnol Bioeng 77:717-722

35.  de Guzman D (2010) Bioplastic development increases with new applications. http://www.icis.com/Articles/2010/10/25/9402443/bioplastic-development-increases-with-new-applications.html webcite. Accessed October 2010

36.  Dorozhkin SV (2009) Calcium orthophosphate-based biocomposites and hybrid biomaterials. J Mater Sci 44:2343-2387

37.  Doug S (2010) Bioplastics: technologies and global markets. BCC research reports PLS050A. http://www.bccresearch.com/report/bioplastics-technologies-markets-pls050a.html webcite

38.  Draget KI (2000) Alginates. In: Philips O, Williams A (eds) Handbook of hydrocolloids, Woodhead, Philadelphia. p 379

39.  Draget KI, Skjåk-Braek G, Smidsrød O (1997) Alginate based new materials. Int J Biol Macromol 21:47-55

40. DSM press release (2008) DSM invests in development of bio-based materials. http://www.observatorioplastico.com/detalle_noticia.php?no_id=73274&seccion=mercado&id_categoria=80002 *webcite*. Accessed March 2008

41. Eichhorn SJ, Baillie CA, Zaferiropouls N, Mwaikambo LY, Ansell MP, Dufresne A, Entwistle KM, Herrera-Franco PJ, Escamilla GC, Groom L, Hughes M, Hill C, Rials TG, Wild PM (2001) Review: current international research into cellulosic fibres and composites. J Material Sc 36:2107-2131

42. Erwin TH, David AG, Jeffrey JK, Robert JW, Ryan PO (2007) The eco-profiles for current and near-future NatureWorks® polylactide (PLA) production. Industrial Biotechnology 3:58-81

43. Eslmai H, Kamal RM (2013) Elongational rheology of biodegradable poly(lactic acid)/poly[(butylene succinate)-co-adipate] binary blends and poly(lactic acid)/poly[(butylene succinate)-co-adipate]/clay ternary nanocomposites. J Appl Polym Sci 127:2290-2306

44. Espigares I, Elvira C, Mano JF, Vlazquez B, Roman JS, Reis RL (2002) New partially degradable and bioactive acrylic bone cements based on starch blends and ceramic fillers. Biomaterials 23(8):1883-1895

45. EurActiv.com (2012) EU calls time on first-generation biofuels. http://www.euractiv.com/climate-environment/eu-signals-generation-biofuels-news-515496*webcite*. Accessed Oct 2012

46. Freier T, Kunze C, Nischan C (2002) In vitro and in vivo degradation studies for development of a biodegradable patch based on poly(3-hydroxybutyrate). Biomaterials 23:2649-2657

47. Fuentes Z, Riquelme MJN, Sánchez-Zapata E, Pérez JAÁ (2010) Resistant starch as functional ingredient: a review. Food Res Int 43:931-942

48. Garlotta D (2001) A literature review of poly (lactic acid). J Polyms and the Envir 9(2):63-84

49. Georg JM, Kristensen M, Astrup A (2012) Effect of alginate supplementation on weight loss in obese subjects completing a 12-week energy restricted diet: a randomized controlled trail. Am J Clin Nutr 96:5-13

50. Goh GH, Heng PWS, Chan LW (2012) Alginates as a useful natural polymer for microencapsulation and therapeutic applications. Carbohydr Polym 88:1-12

51. Gomez-Guille MC, Gimenez B, Lopez CME, Montero MP (2011) Functional bioactive properties of collagen and gelatin from alternative sources: a review. Food Hydrocolloids 25:1813-1827

52. Guangwen C, Shulian L, Fengjun J, Quan Y (2007) Catalytic dehydration of bioethanol to ethylene over $TiO_2/\gamma\text{-}Al_2O_3$ catalyst in microchannel reactors. Catal today 125:111-119

53. Guo M, Liu M, Zhan F, Wu L (2005) Preparation and properties of a slow-release membrane-encapsulated urea fertilizer with superabsorbent and moisture preservation. Ind Eng Chem Res 44:4206-4211

54. Gupta B, Revagade N, Hilborn J (2007) Poly(lactic acid) fiber: an overview. Prog Polym Sci 34:455-482

55. Hafdani FN, Sadeghinia N (2011) A review on applications of chitosan as a natural antimicrobial. World Academy of Sci Engg Technol 50:252-256

56. Hanggi JU (1995) Requirements on bacterial polyesters as future substitute for conventional plastics for consumer goods. FEMS Microbioly Rev 16:213-220

57. Haung YM, Li H, Huang XJ, Hu YC, Hu Y (2008) Advances of bio-ethylene. Chin J Bioprocess Eng 6:1-6

58. Helga E, Svein V (1998) Biosynthesis and applications of alginates. Polym Degradation and Stability 59:85-91

59. Helgerud T, Gaserød O, Fjreide T, Andresen PO, Larsen CK (2009) Alginates. In: Imeson A (ed) Food stabilisers, thickeners and gelling agents, Wiley Blackwell, Oxford. pp 50-72

60. Hirosawa E, Danjo K, Sunada H (2000) Influence of granulating method on physical and mechanical properties, compression behavior, and compactibility of lactose and microcrystalline cellulose granules. Drug Dev Ind Pharm 26:583-593

61. Hori K, Nojiri H, Nonomura M, Okuda F, Yanagida H (2005) Allergen inactivator. US Patent 197319, 20 Nov 2005

62. Huan Z, Joseph GL, Sarit BB (2012) Fabrication aspects of PLA-CaP/PLGA-CaP composites for orthopedic applications: a review. Acta Biomater 8(6):1999-2016

63.    Iain AB, Seal CJ, Wilcox M, Dettmar PW, Pearson PJ (2009) Applications of alginates in food. In: Brend HAR (ed) Alginates: biology and applications. Microbiology monographs 13, Springer, Hiedelberg. pp 211-228

64.    Jane J (1995) Starch properties, modifications and applications. J Macromolecular Sci 32:751-757

65.    Jaspreet S, Lovedeep K, McCarthy OJ (2007) Factors influencing the physico-chemical, morphological, thermal and rheological properties of some chemically modified starches for food applications—a review. Food Hydrocolloids 21:1-22

66.    Jayanth P, Vinod L (2012) Biodegradable nanoparticles for drug and gene delivery to cells and tissue. Adv Drug Deliv Rev 64:61-71

67.    Jeevitha D, Kanchana A (2013) Chitosan/PLA nanoparticles as a novel carrier for the delivery of anthraquinone: synthesis, characterization and *in vitro* cytotoxicity evaluation. Colloids Surf B Biointerfaces 101(1):126-134

68.    Jian-Bing Z, Ling J, Yi-Dong L, Madhusudhan S, Tao L, Yu-Zhong W (2011) Bio-based blends of starch and poly(butylene succinate) with improved miscibility, mechanical properties, and reduced water absorption. Carbohydr Polym 83:762-768

69.    Johnston-Banks FA (1990) Gelatin. In: Harris P (ed) Food gels, Elsevier, London. pp 233-289

70.    Jong ED, Higson A, Walsh P, Maria W (2010) Bio-based chemicals: value added products from biorefineries. IEA Bioenergy Task 42 Biorefinery. 1-34 http://www.iea-bioenergy.task42 biorefineries.com/publications/reports/?eID=dam_frontend_push&docID=2051 *webcite*. Accessed 15 Feb 2012

71.    Jun X, Bao-Hua G (2010) Microbial succinic acid, its polymer poly(butylene succinate), and applications. Microbiology Monographs 14:347-388

72.    Jun X, Bao-Hua G (2010) Poly(butylene succinate) and its copolymers: research, development and industrialization. Biotechnol J 5:1149-1163

73.    Jung S, Lim E, Jong HK (2009) New application of poly(butylene succinate) (PBS) based ionomer as biopolymer: a role of ion group for hydroxyapatite (HAp) crystal formation. J Mater Sci 44:6398-6403

74.  Jung-Ju K, Guang-Zhen J, Hye-Sun Y, Seong-Jun C, Hae-Won K, Ivan BW (2012) Providing osteogenesis conditions to mesenchymal stem cells using bioactive nanocomposite bone scaffolds. Mater Sci Eng C 32:2545-2551

75.  Kamel S, Ali N, Jahangir K, Shah SM, El-Gendy (2008) Pharmaceutical significance of cellulose: a review. Express polymer Letters 2:758-778

76.  Kasirajan S, Ngouajio M (2012) Polyethylene and biodegradable mulches for agricultural applications: a review. Agronomy Sustainable Dev 32(2):501-529

77.  Kathiraser Y, Aroua MK, Ramachandran KB, Tan IKP (2007) Chemical characterization of medium-chain-length polyhydroxyalkanoates (PHAs) recovered by enzymatic treatment and ultrafiltration. J Chem Tech Biotech 82:847-855

78.  Kellomaki M, Niiranen H, Puumanen K, Ashammakhi N, Waris T, Tormala P (2000) Bioabsorbable scaffolds for guided bone regeneration and generation. Biomaterials 21:2495-2505

79.  Kevin JE, Charles MB, John DS, Paul AR, Brian DS, Michael CS, Debra T (2001) Advances in cellulose eater performance and applications. Progress in Polymer Sci 26:1605-1688

80.  Kumano A, Fujiwara N (2008) Cellulose triacetate membranes for reverse osmosis. In: Normam AGF, Li N, Winston Ho WS, Matsuura T (eds) Advanced membrane technology and application, Wiley, New Jersey. pp 21-43

81.  Kumbar SG, Kulkarni AR, Dave AM, Aminabha TM (2001) Encapsulation efficiency and release kinetics of solid and liquid pesticides through urea formaldehyde cross-linked starch, guar gum, and starch + guar gum matrices. J Appli Polym Sci 82:2863-2866

82.  Kunze C, Edgar Bernd H, Androsch R (2006) In vitro and in vivo studies on blends of isotactic and atactic poly (3-hydroxybutyrate) for development of a dura substitute material. Biomaterials 27:192-201

83.  Leathers TD (2003) Biotechnological production and applications of pullulan. Appl Microbiol Biotechnol 62:468-473

84.  Lee SH, Wang S (2006) Biodegradable polymers/bamboo fiber composite with bio-based coupling agent. Compos Part A37:80-91

85.  Lee HC, Anuj S, Lee Y (2001) Biomedical applications of collagen. International J of Pharmaceutics 221:1-22

86.  Lemoigne M (1923) Production d'acide β-oxybutyrique par certaines bact'eries du groupe du Bacillus subtilis. CR. Hebd. Seances Acad. Sci 176:1761

87.  Levison PR (1993) Cellulosics as ion-exchange materials. In: Kennedy JF, Phillips GO, Williams PA (eds) Cellulosics: materials for selective separations and other technologies, Ellis Horwood, Chichester. pp 25-36

88.  Li G, Yong H, Chen C (2011) Discussion on application prospect of starch-based adhesives on architectural gel materials. Adv Materials Res 250:800-803

89.  Li S, Juliane H, Martin KP (2009) Product overview and market projection of emerging biobased products. PRo-BIP 1:1-245

90.  Liu L, Yu J, Cheng L, Qu W (2009) Mechanical properties of poly(butylene succinate) (PBS) biocomposites reinforced with surface modified jute fibre. Composites Part A: Appl Sci Manufacturing 40:669-674

91.  Liu LF, Yu JY, Cheng LD, Yang XJ (2009) Biodegradability of PBS composite reinforced with jute. Polym Degrade Stab 94:90-94

92.  Liu M, Zhang Y, Wu C, Xiong S, Zhou C (2012) Chitosan/halloysite nanotubes bionanocomposites: structure, mechanical properties and biocompatibility. Int J Biological Macromol 51:566-575

93.  Luc A, Eric P (2012) Biodegradable polymers. In: Environmental silicate nano-biocomposites. Green energy and technology. Springer, Hiedelberg. pp 13-39

94.  Luiz A, De Castro R, Morschbacker (2010) A method for the production of one or more olefins, an olefin, and a polymer. US 2010/0069691A1, 18 Mar 2010

95.  Maeda M, Tani S, Sano A, Fujioka K (1999) Microstructure and release characteristics of the minipellet, a collagen based drug delivery system for controlled release of protein drugs. J Controlled Rel 62:313-324

96.  Mainil V, Rahn B, Gogolewski S (1997) Long-term *in vivo* degradation and bone reaction to various polylactides: 1. One-year results. Biomaterials 18:257-266

97. McChalicher CW, Srienc F (2007) Investigating the structure–property relationship of bacterial PHA block copolymers. J Biotechnology 132:296-302

98. Ravi Kumar MNV (2000) A review of chitin and chitosan applications. Reactive Functional Polym 46:1-27

99. Majid J, Elmira AT, Muhammad I, Muriel J, St'ephane D (2010) Poly-lactic acid: production, applications, nanocomposites, and release studies. Comprehensive Rev Food Sci Safety 9(5):552-571

100. Mandel KG, Daggy BP, Brodie DA, Jacoby HI (2000) Review article: alginate-raft formulations in the treatment of heartburn and acid reflux. Aliment Pharmacol Ther 14:669-690

101. Marguerite R (2006) Chitin and chitosan: properties and applications. Progress in Polym Sci 31:603-632

102. Marques AP, Reis RL (2005) Hydroxyapatite reinforcement of different starch-based polymers affect osteoblast-like cells adhesion/spreading and proliferation. Mater Sci Engg 25(2):215-229

103. Maurizio A, Jan JDV, Maria EE, Sabine F, Paolo V, Maria GV (2005) Biodegradable starch/clay nanocomposite films for food packaging applications. Food Chem 93(3):467-474

104. Mendes RL, Reis YP, Bovell AM, Cunha CA, Blitterswijk V, de Bruijn JD (2001) Biocompatibility testing of novel starch-based materials with potential application in orthopedic surgery: a preliminary study. Biomaterials 22:2057-2064

105. Michael C, Dirk C, Harald K, Jan R, Joachim V (2011) Policy paper on bio-based economy in the EU: level playing field for bio-based chemistry and materials. www.bio-based.eu/policy/en webcite. Accessed December 2012

106. Nagarwal RC, Singh PN, Kant S, Maiti P, Pandit JK (2010) Chitosan coated PLA nanoparticles for ophthalmic delivery: characterization, in-vitro and in-vivo study in rabbit eye. J Biomed Nanotechnol 6:648-656

107. Nicolas J, Floriane F, Francoise F, Alan R, Jean PP, Patrick F, Rene SL (2011) Synthesis and properties of poly(butylene succinate): efficiency of different transesterfication catalysts. J Polym Sci Part A: Polym Chem 49:5301-5312

108. Obae HI, Imada K (1999) Morphological effect of microcrystalline cellulose particles on tablet tensile strength. Int J Pharm 182:155-164

109. Onsoyen E (1996) Commercial applications of alginates. Carbohydrates in Europe 14:26-31

110. Orawan S, Uracha R, Pitt S (2008) Electrospun cellulose acetate fiber mats containing asiaticoside or Centella asiatica crude extract and the release characteristics of asiaticoside. Polymer 49(19):4239-4247

111. Ozdemir M, Floros JD (2004) Active food packaging technologies. Crit Rev Food Sci Nutr 44:185-193

112. Papenburg BJ, Liu J, Higuera G, Barradas AMC, Boer J, Blitterswijk VCA, Wessling M, Stamatialis D (2009) Development and analysis of multi-layer scaffolds for tissue engineering. Biomaterials 30:6228-6239

113. Peter WD, Vicki S, Richardson JC (2011) The key role alginates play in health. Food Hydrocolloids 25:263-266

114. Philip S, Keshavarz T, Roy I (2007) Polyhydroxyalkanoates: biodegradable polymers with a range of applications. J Chemical Tech Biotech 2(3):233-247

115. Picker KM, Hoag SW (2002) Characterization of the thermal properties of microcrystalline cellulose by modulated temperature differential scanning calorimetry. J Pharmaceutical Sci 91:342-349

116. Prashant RC, Ishwar BB, Shrikant AS, Rekha SS (2009) Microbial cellulose: fermentive production and applications. Food Technol Biotechnol 47:107-124

117. Qin Y, Cai L, Feng D, Shi B, Liu J, Zhang W, Shen Y (2007) Combined use of chitosan and alginate in the treatment of waste water. J Appl Polym Sci 104:3181-3587

118. Rafael A, Loong TL, Susan EM, Selke HT (2010) Poly(lactic acid): synthesis, structures, properties, processing and applications. Chapter 28:457-467

119. Rajev AJ (2000) The manufacturing techniques of various drug loaded biodegradable poly(lactide-co-glycolide) (PLGA) devices. Biomaterials 21:2475-2490

120. Ramya R, Venkatesan , Jayachanndran Kim S, Sudha PN (2012) Biomedical applications of chitosan: an overview. J Biomaterial Tissue Engg 2:100-111

121. Ravenstijn JTJ (2010) The state-of-the art on bioplastics: products, markets, trends and technologies. Polymedia, Lüdenscheid.

122. Ray SS, Bousmina M (2005) Biodegradable polymers and their layered silicate nanocomposites: in greening the 21st century materials world. Progress Material Sci 50:962-1079

123. Reis KC, Pereira J, Smith AC, Carvalho CWP, Wellner N, Yakimets I (2008) Characterization of polyhydroxybutyrate-hydroxyvalerate (PHB-HV)/maize starch blend films. J Food Engg 89:361-369

124. Rekha MR, Chrndra PS (2007) Pullulan as a promising biomaterial for biomedical applications: a perspective. Trends in Biomaterials and Artificial Organs 20:21-45

125. Roberts GAF (1997) Chitosan production routes and their role in determining the structure and properties of the product. In: Domard M, Roberts AF, Vårum KM (eds) Advances in Chitin Science, vol. 2, National Taiwan Ocean University, Taiwan, Jacques Andre, Lyon. pp 22-31 1998

126. Rubin AL, Stenzel KH, Miyata T, White MJ, Dune M (1973) Collagen as a vehicle for drug delivery: preliminary report. J of Clinical Pharmacology 13:309-312

127. Ruth KG, Hartmann R, Egli T, Zinn M, Ren Q (2007) Efficient production of (R)-3-hydroxycarboxylic acids by biotechnological conversion of polyhydroxyalkanoates and their purification. Biomacromolecules 8:279-286

128. Savenkova L, Gercberga Z, Nikolaeva V, Dzene A, Bibers I, Kalina M (2000) Mechanical properties and biodegradation characteristics of PHB-based films. Process Biochem 35:537-579

129. Schmidmaier G, Wildemann A, Stemberger A, Has MR (2001) Biodegradable poly(D, L-lactide) coating of implants for continuous release of growth factors. J Biomed Mater Res 58:449-455

130. Sena RF, Costelli MC, Gibson LH, Coughlin RW (2006) Enhanced production of pullulan by two strains of A. pullulans with different concentrations of soybean oil in sucrose solution in batch fermentations. Brazilian J Chem Eng 2:507-515

131. Siddhesh NP, Edgar KJ (2012) Alginate derivatization: a review chemistry, properties and applications. Biomaterials 33:3279-3305

132. Singh RS, Saini GK, Kennedy JF (2008) Pullulan: microbial sources, production and applications. Carbohydr Polym 73:515-531

133. Sodian R, Hoerstrup SP, Sperling JS, Daebritz S, Martin DP, Moran AM, Kim BS, Schoen FJ, Vacanti JP, Mayer JE (2000) Early in vivo experience with tissue engineered trileaflet heart valves. Circulation 102:22-29

134. Sodian R, Sperling JS, Martin DP (1999) Tissue engineering of a trileaflet heart valve-early in vitro experiences with a combined polymer. Tissue Engg 5:489-494

135. Steinbüchel A, Valentin HE (1995) Diversity of bacterial polyhydroxyalkanoic acids. FEMS Microbiol Lett 128:219-228

136. Tang H, Ishii D, Mahara A, Murakami S, Yamaoka T, Sudesh K, Samian R, Fujita M, Maeda M, Iwata T (2008) Scaffolds from electrospun polyhydroxyalkanoate copolymers: fabrication, characterization, bio absorption and tissue response. Biomaterials 29:1307-1317

137. Teli MD, Chiplunkar V (1986) Role of thickeners in final performance of reactive prints. Textile Dyer Printer 19:13-19

138. Türesin F, Gürsel I, Hasirci V (2001) Biodegradable polyhydroxyalkanoate implants for osteomyelitis therapy: in vitro antibiotic release. J Biomaterials Sci 12(2):195-207 Polymer Edition

139. Valantin MA, Aubron-Olivier C, Ghosn J, Laglenne E, Pauchard M, Schoen H (2003) Polylactic acid implants to correct facial lipoatrophy in HIV-infected patients: results of the open-label study. Vega Aids 17:2471-2477

140. Valappil S, Misra S, Boccaccini A, Roy I (2006) Biomedical applications of polyhydroxyalkanoates, an overview of animal testing an *in vivo* responses. Expert Rev Med Devices 3:853-868

141. Valerie D, Vinod DV (1998) Pharmaceutical applications of chitosan. Pharmaceutical Sci Technol Today 1:246-253

142. Virginia E, Marie G, Eric P, Luc A (2011) Structure and properties of glycerol plasticized chitosan obtained by mechanical kneading. Carbohydrate Polym 83:947-952

143. Volova T, Shishatskaya E, Sevastianov V, Efremov S, Mogilnaya O (2003) Results of biomedical investigations of PHB and PHB/PHV fibers. Biochem Eng J 16:125-133

144. Vona IA, Costanza JR, Cantor HA, Robert WJ (1965) Manufacture of plastics, vol 1. Wiley, New York. pp 141-142

145. Walle GAM, de Koning GJM, Weusthuis RA, Eggink G (2001) Properties, modifications and applications of biopolyesters. Adv Biochem Eng Biotechnol 71:264-291

146. Wang Z, Itoh Y, Hosaka Y, Kobayashi I, Nakano Y, Maeda I, Umeda F, Yamakawa J, Kawase M, Yagi K (2003) Novel transdermal drug delivery system with polyhydroxyalkanoate and starburst polyamidoamine dendrimer. J Biosci and Bioengg 95(5):541-543

147. Weraporn PA, Sorapong P, Narongchai OC, Ubon I, Puritud J, Sommai PA (2011) Preparation of polymer blends between poly (L-lactic acid), poly (butylene succinate-co-adipate) and poly (butylene adipate-co-terephthalate) for blown film industrial application. Energy Procedia. 9:581-588

148. Westermark S, Juppo AM, Kervinen L, Yliruusi J (1999) Microcrystalline cellulose and its microstructure in pharmaceutical processing. Eur J Pharm Biopharm 48:199-206

149. Williams SF, Martin DP, Horowitz DM, Peoples OP (1999) PHA applications: addressing the price performance issue I. Tissue engineering. Int J Biol Macromolecules 25:111-121

150. Win NN, Stevens WF (2001) Shrimp chitin as substrate for fungal chitin deacetylase. Appl Microbiol Biotechnol 57:334-341

151. Xie ZP, Huang Y, Chen YL, Jia Y (2001) A new gel casting of ceramics by reaction of sodium alginate and calcium iodate at increased temperature. J Mat Sci Lett 20:1255-1257

152. Yan YF, Krishnaiah D, Rajin M, Bono A (2009) Cellulose extraction from palm kernel cake using liquid phase oxidation. J Engg Sci Tech 4:57-68

153. Yang F, Li X, Li G, Zhao N, Zhang X (2002) Study on chitosan and PHBHHx used as nerve regeneration conduit material. J Biomedical Engg 19:25-29

154. Zajic JE, LeDuy A (1973) Flocculant and chemical properties of a polysaccharide from Pullularia pullulans. Appl Microbiol 25:628-635

155. Zeeshan K, Gopiraman M, Yuichi H, Kai W, Kim IS (2013) Cationic-cellulose nanofibers: preparation and dyeability with anionic reactive dyes for apparel application. Carbohydrate 91:434-443

156. Zhao RX, Torley P, Halley PJ (2008) Emerging biodegradable materials: starch- and protein-based bio-nanocomposites. J Mat Sci 43:3058-3071

157. Zhao K, Tian G, Zheng Z, Chen JC, Chen GQ (2003) Production of d-(−)-3-hydroxyalkanoic acid by recombinant *Escherichia coli*. FEMS Microbiol Lett 218:59-64

158. Zhao P, Liu W, Wu Q, Ren J (2010) Preparation, mechanical, and thermal properties of biodegradable polyesters/poly(lactic acid) blends. J of Nanomaterials 2010:1-8

# An Economic and Ecological Perspective of Ethanol Production from Renewable Agro Waste: a Review

Latika Bhatia, Sonia Johri, and Rumana Ahmad

Department of Life Sciences, Institute of Technology & Management University, Gwalior, Madhya Pradesh, 475001, India

## ABSTRACT

Agro-industrial wastes are generated during the industrial processing of agricultural products. These wastes are generated in large amounts throughout the year, and are the most abundant renewable resources on earth. Due to the large availability and composition rich in compounds that could be used in other processes, there is a great interest on the reuse of these wastes, both from economic and environmental viewpoints.

The economic aspect is based on the fact that such wastes may be used as low-cost raw materials for the production of other value-added compounds, with the expectancy of reducing the production costs. The environmental concern is because most of the agro-industrial wastes contain phenolic compounds and/or other compounds of toxic potential; which may cause deterioration of the environment when the waste is discharged to the nature. Although the production of bioethanol offers many benefits, more research is needed in the aspects like feedstock preparation, fermentation technology modification, etc., to make bioethanol more economically viable.

# INTRODUCTION

Since the late 19th century, the mean temperature on earth has increased with 0.8°C and the major part of this increase is likely due to anthropogenic emissions of greenhouse gases. Carbon dioxide is the type of greenhouse gas with largest emission and this originates from the combustion of fossil fuels as coal, oil and natural gas (Sun and Cheng 2002). In USA, transportation accounts for 30% of the total energy consumption. Burning fossil fuels such as coal and oil releases $CO_2$, which is a major cause of global warming. With only 4.5% of the world's population, the US is responsible for about 25% of global energy consumption and 25% of global $CO_2$ emissions. The average price of gasoline in 2005 was $2.56 per gallon, which was $0.67 higher than the average price of gasoline in the previous year. Yet in June 2008, the average price of gasoline in the US reached $4.10 per gallon (Kumar et al. 2009).

Soaring oil prices associated with concerns of climate change and national energy security are driving us to utilize sustainable alternative energy sources, such as solar energy, nuclear energy, wind energy, hydropower, tidal energy, and so on (Lynd et al. 2003). With the inevitable depletion of world's energy supply, there has been an increasing interest worldwide in alternative sources of energy. Unlike fossil fuels, ethanol is a renewable energy source produced through fermentation of sugars and used as a partial gasoline replacement in a few countries of the world (Sharma et al.2007).

Bioethanol market is expected to reach $100 \times 10^9$ liters in 2015. The largest producers in the world are the US, Brazil, and China. In 2009,

US produced 39.5x10$^9$ liters of ethanol using corn as a feedstock while the second largest producer, Brazil, created about 30x10$^9$ liters of ethanol using sugarcane. China is a country that has invested much in the production of ethanol, and is nowadays one of the largest ethanol producers (Ivanova et al. 2011).

Ethanol contains 35% oxygen, which results in a complete combustion of fuel and thus lowers the emission of harmful gases. Moreover, ethanol production uses energy from renewable sources only; hence, no net carbon dioxide is added to the environment, thus reducing green-house gas emissions. It has also been well established now that ethanol increases the octane number, decreases the Reid vapor pressure and produces fuel with clean burning characteristics (Dhillon et al. 2007). Moreover, neat (unblended) ethanol can be burned with greater efficiency, and is thought to produce smaller amounts of ozone precursors (thus decreasing urban air pollution), and is particularly beneficial with respect to low net $CO_2$ put into the atmosphere.

The increasing demand for various industrial purposes such as alternative source of energy, industrial solvents, cleansing agents and preservatives, has necessitated increased production of ethanol (Brooks 2008). Furthermore, ethanol by fermentation offers a more favorable trade balance, enhanced energy security, and a major new crop for a depressed agricultural economy. Ethanol is considerably less toxic to humans than is gasoline (or methanol). Ethanol also reduces smog formation because of low volatility; its photochemical reactivity and low production of combustion products. Furthermore, low levels of smog-producing compounds are formed by its combustion (Wyman and Hinman 1990). In addition, the low flame temperature of ethanol results in good engine performance.

Currently, bioethanol is being commercially produced only from edible feedstock such as corn-starch and sugarcane juice. The European Union (EU) had established a goal of 5.75% biomass-derived transportation fuels by December, 2010. The use of fuel ethanol has been quite successful in Brazil, where it is being produced at a very low cost by fermentation of sugarcane. In the US, corn is the dominant biomass feedstock for production of ethanol, and in the EU, straw and other agricultural wastes are the preferred types of biomass for ethanol production (Raposo et al. 2009). These bioethanol production systems pose a concern about competition with food and feed supplies. To avoid this competition, bioethanol production from non-edible lignocellulosic biomass such as wheat straw, rice straw, bagasse, corn stover, wood, peels of fruits and vegetables is attracting keen interest. The current production and use of bioethanol processes are a starting point. It is our belief that the next generational change in the use of bioresources will come from a total integration of innovative plant resources, synthesis of biomaterials, and generation of biofuels and biopower (Figure1).

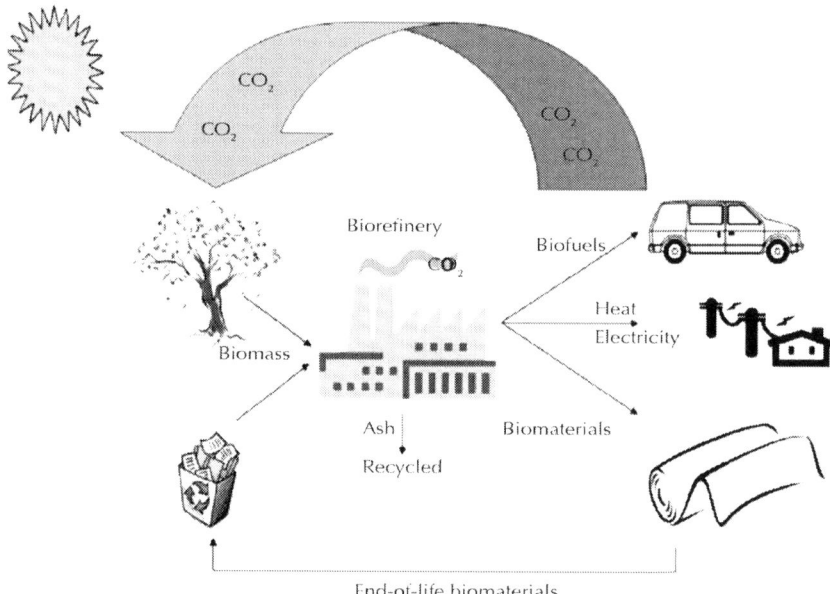

**Figure 1:** The fully integrated agro-biofuel-biomaterial-biopower cycle for sustainable technologies. (Ragauskas et. al Ragauskas et al.2006).

The present review is a concise overview of current and latest developments in ethanol production with special emphasis on the choice of lignocellulosic substrates, pretreatment methods and types of microorganisms that have been used for optimal, ecological and economic production of ethanol. Also reviewed are the different fungal and bacterial lignocellulolytic enzymatic systems including the current status of the technology for bioconversion of lignocellulose residues by microorganisms (particularly yeasts and fungi), with focus on the most economical and eco-friendly method for ethanol production.

# Lignocellulosic Biomass

Lignocellulose is a renewable organic material and is the major structural component of all plants. Lignocellulose consists of three major components:

- Cellulose, the major constituent of all plant material and the most abundant organic molecule on earth, is a linear biopolymer of anhydroglucopyranose-molecules, connected by $\beta$-1, 4-glycosidic bonds. Cellulose or $\beta$-1-4-glucan is a linear polysaccharide polymer of glucose made of cellobiose units. The cellulose chains are packed by hydrogen bonds in so-called 'elementary microfibrils'. These fibrils are attached to each other by hemicelluloses, amorphous polymers of different sugars as well as other polymers such as pectin, and covered by lignin. The microfibrils are often associated in the form of bundles or macrofibrils. This special and complicated structure makes cellulose resistant to both biological and chemical treatments. (Delmer and Amor 1995, Morohoshi 1991, Ha et al. 1998).

- Hemicellulose, the second most abundant component of lignocellulosic biomass, is a heterogeneous polymer of pentoses (including xylose and arabinose), hexoses (mainly mannose, less glucose and galactose) and sugar acids. Hemicellulose is less complex, its concentration in lignocellulosic biomass is 25 to 35% and it is easily hydrolysable to fermentable sugars (Saha et al.2007). The dominant sugars in hemicelluloses are mannose in softwoods and xylose in hardwoods and agriculture residues (Persson et al. 2006, Lavarack et al. 2002, Balan et al. 2009).

- Lignin, the third main heterogeneous polymer in lignocellulosic residues, generally contains three aromatic alcohols including coniferyl alcohol, sinapyl and p-coumaryl. Lignin serves as a sort of 'glue' giving the biomass fibers its structural strength. Lignin acts as a barrier for any solutions or enzymes by linking to both hemicelluloses and cellulose and prevents penetration of lignocellulolytic enzymes to the interior lignocellulosic structure. Not surprisingly, lignin is the most recalcitrant component of lignocellulosic material to degrade (Zaldivar et al. 2001, Hamelinck et al. 2005).

# Lignocellulose Substrates Used for Ethanol Production

Sweet sorghum bagasse can be converted efficiently into fermentable sugars (and is a new potential raw material for fuel ethanol production) by $SO_2$ catalyzed steam pretreatment at 190°C for 10 min or 200°C for 5 min followed by enzymatic hydrolysis with a result of 89-92% glucan conversion (Sipos et al. 2009). Hemp and ensiled hemp can be converted into ethanol with steam pretreatment (2% $SO_2$ catalyst, 210°C for 5 min) followed by simultaneous saccharification and fermentation at high solid loading (7.5% water insoluble solids[WIS]) with a result of 171–163 g ethanol/kg raw material (Sipos et al. 2010). In Brazil, ethanol is usually produced from cane juice, whereas in USA, starch-crops such as corn are usually used for ethanol production (Sanchez 2009). Using sugars or corn as the main source for ethanol production caused a great deal of controversy due to its effect on food production and costs, which has made it difficult for ethanol to become cost competitive with fossil fuels. These concerns became a driving force in the generation of new biofuel research using lignocellulosic wastes produced by many different industries. Apart from corn and cane juice; wheat, oat and barley straw has also been routinely used to produce up to 0.52 million gallons of ethanol per year (Hahn et al. 2006). China is the world's largest sweet potato (*Ipomoea batatas* Lam.) producer (accounting for 85% of global production), with the output exceeded 100 M tons in 2005 (Lu et al.2006). Zhang et al. (2011) reported sweet potato as an attractive feedstock for bioethanol production from both economic and environment friendly standpoints.

Lignocellulosic wastes are produced in large amounts by different industries including forestry, pulp and paper, agriculture and food, in addition to different wastes from municipal solid waste (MSW), and animal wastes (Sims 2003, Kim and Dale 2004, Kalogo et al. 2007, Champagne 2007, Wen et al. 2004). Those derived from agricultural activities include materials such as straw, stem, stalk, leaves, husk, shell, peel, lint, seed/stones, pulp or stubble from fruits, legumes or cereals (rice, wheat, corn, sorghum, barley), bagasses generated from sugarcane or sweet sorghum milling, spent coffee grounds, brewer's spent grains, and many others. These potentially valuable materials were treated as waste in many countries in the past, and still are today in some developing countries, which raises many environmental concerns (Palacios-Orueta et al. 2005). Significant efforts, many of which have been successful, have been made to convert these lignocellulosic residues to valuable products such as biofuels, chemicals and animal feed (Howard et al. 2003). Banana peel, an agro waste can be used as a substrate for ethanol production owing to its rich carbohydrate, crude proteins and reducing sugars. Moreover, banana peels are affordable and renewable low cost raw material which makes it potential feedstock for ethanol production (Bhatia and Paliwal 2010). Similarly pineapple is the second harvest of importance after bananas, contributing to over 20% of the world production of tropical fruits (Coveca 2002). Thailand, Philippines, Brazil and China are the main pineapple producers in the world supplying nearly 50% of the total output. Other important producers include India, Nigeria, Kenya, Indonesia, México and Costa Rica and these countries provide most of the remaining fruit available (50%). Isitua and Ibeh 2010 assayed the feasibility of obtaining ethanol from pineapple waste with the purpose of obtaining a valuable product from the residues of the juice and canning industries.

Large volume of bagasse is generated during sugarcane processing. Agricultural profitability and environmental protection issues are associated with disposal of bagasse. In recent years, potential efforts have been directed towards the utilization of cheap renewable agricultural resources, such as sugarcane bagasse as alternative substrate for ethanol production (Bhatia and Paliwal 2011). Rice is the major crop grown worldwide with an annual productivity around 800 million metric tonnes that corresponds with the large production of rice straw. In search for viable alternatives of biofuels, paddy straw has been pursued as suitable lignocellulosic waste for ethanol production (Wati et al.2007).

Feasibility of lignocellulosic material for ethanol production has been explored around the world depending upon availability. Production of ethanol from wheat straw, one of the most abundant agricultural wastes, has been extensively studied (Ballesteros et al. 2004, Curreli et al. 2002, Curreli et al. 1997, Talebnia et al. 2010). The average yield of wheat straw is 1.3–1.4 lb per lb of wheat grain (Montane et al. 1998). According to Ballesteros et al. 2006, under the 60% ground cover practice, about 354 millions of tons of wheat straw could be available globally and could produce 104 GL of bioethanol. Europe production would account for about 38% of this world bioethanol capacity. In Spain, grain industry generates important amounts of wheat straw, a part of which is used as bedding straw and the remainder is burned or left on the land to fertilize the soil. Bioconversion of this residue to fuel ethanol would provide an attractive possibility to boost the development of biofuels in a sustainable way.

## Overview of Lignocellulosic Fermentation

Schematic picture for the conversion of lignocellulosic biomass to ethanol, including the major steps can been seen in Figure 2. Pretreatment of the lignocellulosic residues is necessary because hydrolysis of non-pretreated materials is slow, and results in low product yield. Some pretreatment methods increase the pore size and reduce the crystallinity of cellulose (Dawson and Boopathy 2007). Pretreatment also makes cellulose more accessible to the cellulolytic enzymes, which in return reduces enzyme requirements and, thus, the cost of ethanol production. The pretreatment not only enhance the bio-digestibility of the wastes for ethanol production, but also results in enrichment of the difficult biodegradable materials, and improves the yield of ethanol from the wastes.

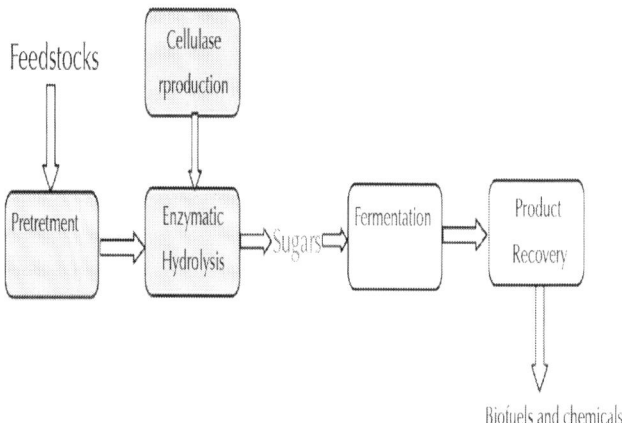

**Figure 2:** Major steps involved in the conversion of lignocellulosic biomass to ethanol (Dashtban et al.2009).

Post pretreatment, the recalcitrant lignocellulosic biomass becomes susceptible to acid and/or enzymatic hydrolysis as the cellulosic microfibrils are exposed and/or accessible to hydrolyzing agents (Jacobsen and Wyman 2000). In the pretreatment process, small amounts of cellulose and most of hemicellulose is hydrolyzed to sugar monomers; mainly D-xylose and D-arabinose. The pretreated biomass is then subjected to filtration to separate liquids (hemicellulose hydrolysate) and solid (lignin and cellulose). After detoxification, the liquid is sent to a xylose (pentose) fermentation column for ethanol production. Solids are subjected to hydrolysis (also called second stage hydrolysis). This process is mainly accomplished by enzymatic methods using cellulases. Mild acid hydrolysis using sulfuric and hydrochloric acids is an alternative procedure (Zhang and Lynd 2004). The hydrolyzed sugars such as D-glucose, D-galactose, and D-mannose, can be readily fermented to ethanol using various strains of *Saccharomyces cerevisae*. The pentoses (D-xylose and D-arabinose) from hemicellulose hydrolysis are not easily utilized by *Saccharomyces* strains; therefore, genetically modified strains of *Pichia stipitis*, *Zymomonas mobilis*, are used for their fermentation. *Candida shehatae* is capable of co-fermenting both pentoses and hexoses to ethanol and other value-added products at high yields (Betancur 2005, Senthilkumar and Gunasekaran 2005).

Numerous pretreatment strategies have been developed to enhance the reactivity of cellulose and to increase the yield of fermentable sugars. Typical goals of pretreatment include:

- Production of highly digestible solids that enhances sugar yields during enzyme hydrolysis, avoidance of degradation of sugars (mainly pentoses) including those derived from hemicelluloses.

- Minimization of formation of inhibitors for subsequent fermentation steps.

- Recovery of lignin for conversion into valuable co-products.

- Cost effectiveness by operating in reactors of moderate size and by minimizing heat and power requirements (Mosier et al. 2005, Sun and Cheng 2007, Yang and Wyman 2008). Figure 3 depicts schematic of goals of pretreatment on lignocellulosic material.

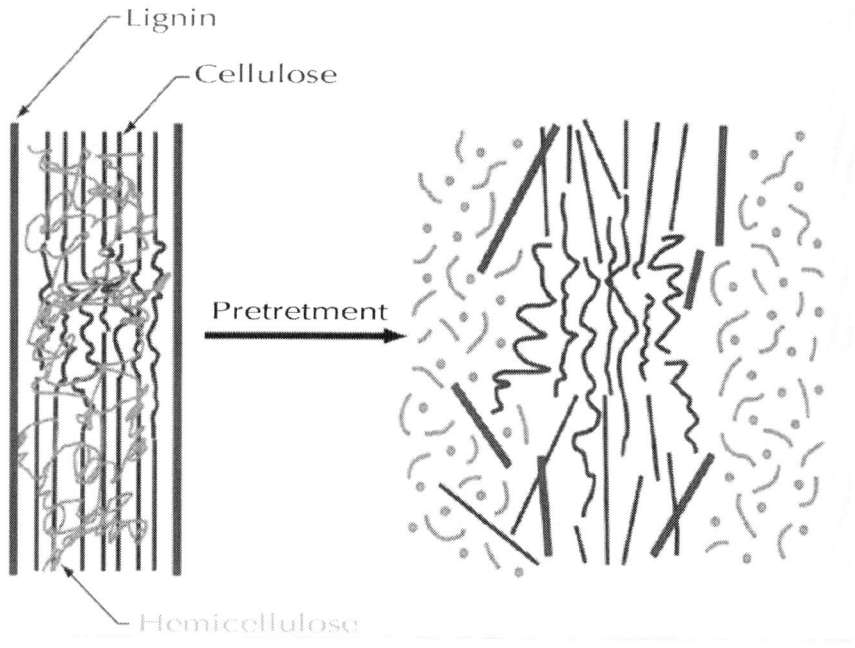

**Figure 3:** Schematic of goals of pretreatment on lignocellulosic material

# Physical Pretreatments Methods

Physical pretreatments methods such as ball milling and grinding have been used for degradation of lignocelluloses with limited success. This method of pretreatment being cost effective and ecofriendly, and one on which relatively little work has been done and reported, so far, would form one of the thrust areas of future research.

Waste materials can be comminuted by a combination of chipping, grinding and milling to reduce cellulose crystallinity. The size of the materials is usually 10–30 mm after chipping and 0.2–2 mm after milling or grinding. Vibratory ball milling has been found to be more effective in breaking down the cellulose crystallinity of spruce and aspen chips and improving the digestibility of the biomass than ordinary ball milling (Millet et al. 1976). The power requirement of mechanical comminution of agricultural materials depends on the final particle size and the waste biomass characteristics (Cadoche and Lopez 1989).

Pyrolysis has also been used for pretreatment of lignocellulosic materials. When the materials are treated at temperatures greater than 300°C, cellulose rapidly decomposes to produce gaseous products and residual char (Kilzer and Broido 1965, Shafizadeh and Bradbury 1979). The decomposition is much slower and less volatile products are formed at lower temperatures.

The efficiency of ultrasound in the processing of vegetal materials has been already proved (Vinatoru et al. 1999). The known ultrasounds benefits, such as swelling of vegetal cells and fragmentation due to the cavitational effect associated with the ultrasonic treatment, act by increasing the yield and by shortening of the extraction time. The effect of ultrasound on lignocellulosic biomass has been employed in order to improve the extractability of hemicelluloses (Ebringerova et al. 2002), cellulose (Pappas et al. 2002), lignin (Sun and Tomkinson 2002) or to get clean cellulosic fiber from used paper (Scott and Gerber 1995) but only few attempts to improve the susceptibility of lignocellulosic materials to biodegradation by using ultrasound have been described. It was found out that ultrasound has a beneficial effect on saccharification processes (Rolz 1986). Sonication has been reported to decrease cellulase requirements by 1/3 to 1/2 and to increase ethanol production from mixed waste office paper by approximately 20% (Wood et al. 1997). It was notice that the effect of ultrasound fragmentation of Avicel (microcrystalline cellulose formed by acid treatment) is similar to that of the enzymes for short incubation intervals (Gama et al. 1997). The time needed for ultrasonic treatment could be reduced when increasing the irradiation power (Imai et al. 2004).

# Chemical Pretreatment Methods

## *Alkaline Pretreatment*

Alkaline pretreatment involves the use of bases, such as sodium, potassium, calcium, and ammonium hydroxide, for the pretreatment of lignocellulosic biomass. The use of an alkali causes the degradation of ester and glycosidic side chains resulting in structural alteration of lignin, cellulose swelling, partial decrystallization of cellulose (Cheng et al. 2010, Ibrahim et al. 2011, McIntosh et al.2010) and partial solvation of hemicelluloses (McIntosh et al. 2010, Sills et al. 2011). Sodium hydroxide has been extensively studied for many years, and it has been shown to disrupt the lignin structure of the biomass, increasing the accessibility of enzymes to cellulose and hemicellulose (MacDonald et al. 1983, Soto et al. 1994, Zhao et al., 2008). Another alkali that has been used for the pretreatment of biomass is lime. Lignocellulosic feedstocks that have been shown to benefit from this method of pretreatment are corn stover, switchgrass, bagasse, wheat, and rice straw.

The conditions for alkaline pretreatment are usually less severe than other pretreatments. It can be performed at ambient conditions, but longer pretreatment times are required than at higher temperatures. The advantage of lime pretreatment is that the cost of lime required for pretreatment of a given quantity of biomass is lowest among alkaline treatments. Most commonly used alkali in the alkali pretreatment processes are NaOH and $Ca(OH)_2$. This process results in (i) the removal of all lignin and part of hemicellulose, and (ii) increased reactivity of cellulose in further hydrolysis steps (Hamelinck et al. 2005), especially, enzymatic hydrolysis. Effective removal of lignin minimizes adsorption of enzyme onto lignin and thus allows for effective interactions with cellulose (Aswathy et al. 2010). Between NaOH and $Ca(OH)_2$, pretreatment with $Ca(OH)_2$ is preferable because it is less expensive, more safer as compared to NaOH and it can be easily recovered from the hydrolysate by reaction with $CO_2$ (Mosier et al. 2005).

## Acid Pretreatment Methods

Acid pretreatment involves the use of concentrated and diluted acids to break the rigid structure of the lignocellulosic material. The most commonly used acid is dilute sulphuric acid ($H_2SO_4$), which has been commercially used to pre-treat a wide variety of biomass types-switchgrass, corn stover, spruce (softwood), and poplar. Acid pretreatment (removal of hemicellulose) followed by alkali pretreatment (removal of lignin) has shown to yield relatively pure cellulose (Wingren et al. 2003, Taherzadeh and Karimi 2008). Strong acid allows complete breakdown of the components in the biomass to sugars, but also requires large volumes of concentrated sulfuric acid and can result in the production of furfural, an inhibitory byproduct (Goldstein and Easter 1992). Dilute acid allows reduced acid concentrations, but requires higher temperatures, and again gives furfural. A key advantage of acid pretreatment is that a subsequent enzymatic hydrolysis step is sometimes not required, as the acid itself hydrolyses the biomass to yield fermentable sugars (Zhu et al. 2009). A mixture of $H_2SO_4$ and acetic acid resulted in 90% saccharification (DeMoraes-Racha et al. 2010). Hemicellulose and lignin are solubilized with minimal degradation, and the hemicellulose is converted to sugars with acid pretreatment. The major drawback to these acid processes is the cost of acid and the requirement to neutralize the acid after treatment.

# Wet Oxidation

Wet oxidation utilizes oxygen as an oxidizer for compounds dissolved in water. Typically, the procedure for wet oxidation consists of drying and milling lignocellulosic biomass to obtain particles that are 2 mm in length, to which water is added at a ratio of 1 L to 6 g biomass. Wet oxidation has been used to fractionate lignocellulosic material by solubilizing hemicellulose and removing lignin (Martin et al. 2007, Banerjee et al. 2009). It has been shown to be effective in pretreating a variety of biomass such as wheat straw, corn stover, sugarcane bagasse, cassava, peanuts, rye, canola, faba beans, and reed to obtain glucose and xylose after enzymatic hydrolysis (Martin et al. 2008, Banerjee et al. 2009, Ruffell et al. 2010, Szijarto et al. 2009, Martin and Thomsen 2007). During wet oxidation, lignin is decomposed to carbon dioxide, water and carboxylic acids. Biomass such as straw, reed and other cereal crop residues have a dense wax coating containing silica and protein which is removed by wet oxidation (Schmidt et al. 2002).

Wet oxidation has been combined with other pretreatment methods to further increase the yield of sugars after enzymatic hydrolysis. Combining wet oxidation with alkaline pretreatment has been shown to reduce the formation of byproducts, thereby decreasing inhibition. Bjerre et al. ( 1996) used wet oxidation and alkaline hydrolysis of wheat straw (20 g straw/l, 170°C, 5–10 min), and achieved 85% conversion yield of cellulose to glucose. Wet oxidation combined with base addition readily oxidizes lignin from wheat straw, thus making the polysaccharides more susceptible to enzymatic hydrolysis. Furfural and hydroxymethylfurfural, known inhibitors of microbial growth when other pretreatment systems are applied, were not observed following the wet oxidation treatment (Azzam 1989).

# Green Solvents

Processing of lignocellulosic biomass with ionic liquids (IL) and other solvents has gained importance in the last decade due to the tunability of the solvent chemistry and hence the ability to dissolve a wide variety of biomass types. Ionic liquid (IL) was found to possess a great potential in dissolving cellulose (Swatloski et al. 2002). Ionic liquids are salts, typically composed of a small anion and a large organic cation, which exist as liquids at room temperature and have very low vapor pressure. The chemistry of the anion and cation has been tuned to generate a wide variety of liquids which can dissolve a number of biomass types- corn stover (Cao et al. 2010), cotton (Zhao et al. 2009), bagasse (Wang et al. 2009), switchgrass, wheat straw (Li et al. 2009).

Dadi and coworkers (2007) have studied the enzymatic hydrolysis of Avicel regenerated from two different ILs, 1-*n*-butyl-3-methylimidazolium chloride and 1-allyl-3-methylimidazoliumchloride. Hydrolysis kinetics of the IL-treated cellulose was significantly enhanced. A limitation in using ionic liquids is the fact they tend to inactivate cellulose.

A solvent which has been effective in dissolution of cellulose and has a low vapor pressure similar to that of the ionic liquids is N-methyl morpholine N-oxide (NMMO). NMMO retains all the advantages of the ionic liquids ability to dissolve a variety of lignocellulosic substrates (Kuo and Lee 2009, Shafiei et al. 2010) without the need to chemically modify them and >99% of the solvent can be recovered due to its low vapor pressure (Perepelkin 2007). It is also nontoxic and biodegradable as proven by the work of Lenzing and other researchers (Rosenau et al. 2001). Further research is needed to evaluate and improve the economics of usage of ILs and NMMO for pretreatment of biomass. Pretreatment of lignocellulosic materials with acidified organic solvents (mixture of 80% ethylene glycol, 19.5% water and 0.5% HCl at 178°C for 90 min) has also been successfully used (Yamashita et al. 2010). The advantages of these methods include recovery and recycling of organic solvents as they can be easily distilled out. The disadvantages are that the process requires expensive high pressure equipment. Their performances could be improved by heating, microwave, or sonication (ElSeoud et al. 2007, Zhu et al. 2006a).

# Physicochemical Pretreatment Methods

## *Steam-Explosion*

Steam-Explosion pretreatment is one of the most commonly used pretreatment options, as it uses both chemical and physical techniques in order to break the structure of the lignocellulosic material (McMillan 1994). This hydrothermal pretreatment method subjects the material to high pressures and temperatures for a short duration of time after which it rapidly depressurizes the system, disrupting the structure of cellulose microfibrils. The disruption of the fibrils increases the accessibility of the cellulose to the enzymes during hydrolysis.

Steam explosion is typically initiated at a temperature of 160–260°C (corresponding pressure 0.69–4.83 MPa) for several seconds to a few minutes before the material is exposed to atmospheric pressure. The process causes hemicellulose degradation and lignin transformation due to high temperature, thus increasing the potential of cellulose hydrolysis (Ballesteros et al. 2006, Chornet et al. 1988, Focher et al. 1991).

However, some disadvantages have been seen when using this process. Dilute acids are required to be added during softwood pretreatment or even when increased yields are warranted for lower acetylated feedstock. The factors that affect steam explosion pretreatment are residence time, temperature, chip size and moisture content (Duff and Murray 1996). Recent studies indicate that lower temperature and longer residence time are more favorable (Wright 1998).

# Liquid Hot Water (LHW)

Much like the steam-explosion process, liquid hot water (LHW) pretreatment uses water at elevated temperatures and high pressures to maintain its liquid form in order to promote disintegration and separation of the lignocellulosic matrix. Temperatures can range from 160°C to 240°Cover lengths of time ranging from a few minutes up to an hour with temperatures dominating the types of sugar formation and time dominating the amount of sugar formation (Yu et al. 2010).

This process has been found to be advantageous from a cost standpoint in that no additives such as acid catalysts are required. Furthermore, expensive reactor systems have not been necessary to use due to the low corrosive nature of this pretreatment technique. Neutralization of degradation products is not needed due to their fractionation and utilization in the liquid fraction. In the same sense, inhibitory products have not been reported to form overwhelmingly in the respective fractions allowing higher yields under specific conditions.

# Ammonia Fiber Explosion (AFEX)

The ammonia fiber/freeze explosion (AFEX) process is another physicochemical process, much like steam explosion pretreatment, in which the biomass material is subjected to liquid anhydrous ammonia under high pressures and moderate temperatures and is then rapidly depressurized. The moderate temperatures (60°C to 100°C) are significantly less than that of the steam explosion process, thus allowing less energy input and overall cost reduction associated with the process (Alizadeh et al. 2005, Teymouri et al. 2004, Chundawat et al. 2007).

There have been extensive literature reviews on this type of pretreatment over the last decade, focusing on the advantages and disadvantages of the AFEX process used for different feedstocks (Sun and Cheng 2002, Mosier 2005). An overview of some of the advantages include lower moisture content, lower formation of sugar degradation products due to moderate conditions, 100% recovery of solid material, and the ability for ammonia to lessen lignin's effect on enzymatic hydrolysis. A smaller number of disadvantages can be seen in the form of higher costs due to recycle and treatment of chemicals that are being used.

# Ammonia Recycle Percolation (ARP)

Ammonia recycle percolation (ARP) has been paired with the AFEX pretreatment process by many authors, but it can have some different characteristics that need to be taken into consideration when looking at different pretreatment options (Kim and Lee 2005). In this process, aqueous ammonia of concentration between 5-15% (wt %) is sent through a packed bed reactor containing the biomass feedstock at a rate of about 5 ml/min. The advantage with this process over AFEX is its ability to remove a majority of the lignin (75–85%) and solubilize more than half of the hemicellulose (50–60%) while maintaining high cellulose content (Kim and Lee 2005). Primarily, herbaceous biomass has been most treated with this process: 60-80% delignification has been achieved for corn stover and 65–85% delignification for switchgrass (Iyer et al. 1996).

# Supercritical Fluid (SCF) Pretreatment

A supercritical fluid is a material which can be either liquid or gas, used in a state above the critical temperature and critical pressure where gases and liquids can coexist. It shows unique properties that are different from those of either gases or liquids under standard conditions-it possesses a liquid like density and exhibits gas-like transport properties of diffusivity and viscosity (King and Srinivas2009). Thus, SCF has the ability to penetrate the crystalline structure of lignocellulosic biomass overcoming the mass transfer limitations encountered in other pretreatments. The lower temperatures used in the process aids in the stability of the sugars and prevents degradation observed in other pretreatments. Kim and Hong 2001 investigated supercritical $CO_2$ pretreatment of hardwood (Aspen) and southern yellow pine with varying moisture contents followed by enzymatic hydrolysis. SCF pretreatment showed significant enhancements in sugar yields when compared to thermal pretreatments without supercritical $CO_2$. Alinia and coworkers ( 2010) investigated the effect of pretreatment of dry and wet wheat straw by supercritical $CO_2$ alone and by a combination of $CO_2$ and steam under different operating conditions (temperature and residence time in the reactors). It was found that a combination of supercritical $CO_2$ and steam gave the best overall yield of sugars.

# Biological Pretreatment Methods

Biological pretreatment uses microorganisms and their enzymes selectively for delignification of lignocellulosic residues and has the advantages of a low-energy demand, minimal waste production and a lack of environmental effects. In biological pretreatment processes, microorganisms such as brown-, white- and soft-rot fungi are used to degrade lignin and hemicellulose in waste materials Schurz (1978). White-rot basidiomycetes possess the capabilities to attack lignin. *Penicillium chrysosporium*, for example, has been shown to non-selectively attack lignin and carbohydrate (Anderson and Akin 2008). *P. chrysosporium* has been successfully used for biological pretreatment of cotton stalks by solid state cultivation (SSC) and results have shown that the fungus facilitates the conversion into ethanol (Shi et al., 2008). Brown rots mainly attack cellulose, while white and soft rots attack both cellulose and lignin. White-rot fungi are the most effective basidiomycetes for biological pretreatment of lignocellulosic materials (Fan et al. 1987). Other basidiomycetes such as*Phlebia radiata*, *P. floridensis* and *Daedalea flavida*, selectively degrade lignin in wheat straw and are good choices for delignification of lignocellulosic residues (Arora and Chander 2002). *Ceriporiopsis subvermispora*, however, lacks cellulases (cellobiohydrolase activity) but produces manganese peroxide and laccase, and selectively delignifies several different wood species (Ferraz 2003). The advantages of biological pretreatment include low energy requirement and mild environmental conditions. However, the rate of hydrolysis in most biological pretreatment processes is very low.

# Hydrolysis of Pretreated Biomass

After pretreatment, the released cellulose and hemicelluloses are hydrolyzed to soluble monomeric sugars (hexoses and pentoses) using cellulases and hemicellulases, respectively. The initial conversion of biomass into sugars is a key bottleneck in the process of biofuel production and new biotechnological solutions are needed to improve their efficiency, which would lower the overall cost of bioethanol production. Enzymatic hydrolysis has been considered key to cost-effective bioethanol in the long run, and the reaction is carried out with mainly cellulase and hemicellulase for cellulose and hemicellulose, respectively. The advantages of using enzyme (cellulase) over acid is to eliminate corrosion problems and lower maintenance costs with mild processing conditions to give high yields.

Despite the fact that some fungal strains have the advantages of being thermostable and producing cellulases, most of these fungal strains do not produce sufficient amounts of one or more lignocellulolytic enzymes required for efficient bioconversion of lignocellulosic residues to fermentable sugars. In addition, plant cell walls are naturally resistant to microbial and enzymatic (fungal and bacterial) deconstruction, collectively known as 'biomass recalcitrance' (Himmel et al. 2007). These rate-limiting steps in the bioconversion of lignocellulosic residues to ethanol remain one of the most significant hurdles to producing economically feasible cellulosic ethanol. Improving fungal hydrolytic activity and finding stable enzymes capable of tolerating extreme conditions has become a priority in many recent studies.

# Fungal Extracellular Cellulases

Enzymatic saccharification of lignocellulosic materials such as sugarcane bagasse, corncob, rice straw,*Prosopis juliflora, Lantana camara*, switch grass, saw dust, and forest residues by cellulases for biofuel production is perhaps the most popular application currently being investigated (Kuhad et al.2010, Sukumaran et al. 2005). Both bacteria and fungi can produce glucanases (cellulases) that hydrolyze of lignocellulosic materials. These microorganisms can be aerobic or anaerobic and mesophilic or thermophilic. Bacteria belonging to genera of *Clostridium, Cellulomonas, Bacillus,Thermomonospora, Ruminococcus, Bacteriodes, Erwinia, Acetovibrio, Microbispora,* and *Streptomyces*are known to produce cellulase (Bisaria 1998). Anaerobic bacterial species such as *Clostridium phytofermentans, Clostridium thermocellum, Clostridium hungatei, and Clostridium papyrosolvens*produces cellulases with high specific activity (Duff and Murray 1996, Bisaria 1998). Most commercial glucanases (cellulases) are produced by *Trichoderma ressei* and β-D-glucosidase is produced from*Aspergillus niger* (Kaur et al. 2007). Fungi known to produce cellulases include *Sclerotium rolfsii,Phanerochaete chrysosporium* and various species of *Trichoderma, Aspergillus, Schizophyllum and Penicillium* (Sternberg 1976, Fan et al. 1987, Duff and Murray 1996). Among the fungi, *Trichoderma*species have been extensively studied for cellulase production (Sternberg 1976).

High temperature and low pH tolerant enzymes are preferred for the hydrolysis due to the fact that most current pretreatment strategies rely on acid and heat (Turner et al. 2007). In addition, thermostable enzymes have several advantages including higher specific activity and higher stability which improve the overall hydrolytic performance (Viikari et al. 2007). Ultimately, improvement in catalytic efficiencies of enzymes will reduce the cost of hydrolysis by enabling lower enzyme dosages. Some fungal strains such as *T. emersonii* (Grassick et al. 2004), *Chaetomium thermophilum* (Li et al.2006) and *Corynascus thermophilus* (Rosgaard et al. 2006) can produce thermostable enzymes which are stable and active at elevated temperatures (60°C) well above their optimum growth temperature (30-55°C) (Maheshwari et al., 2000). Due to the promising thermostability and acidic tolerance of thermophilic fungal enzymes, they have good potential to be used for hydrolysis of lignocellulosic residues at industrial scales.

The anaerobic bacteria *Clostridium thermocellum* and *Clostridium cellulovorans* and the filamentous fungus *Trichoderma reesei* are well known as strongly cellulolytic and xylanolytic microorganisms. *C.thermocellum* and *C. cellulovorans* produce a cellulosome complex consisting of cellulase and hemicellulase organized on the cell surface (Demain et al. 2005); *T. reesei*, meanwhile, extracellularly secretes three types of cellulolytic enzyme, including five endoglucanases (EG [EC 3.2.1.4]) (Pere et al. 2001, Dienes et al. 2004), two cellobiohydrolases (CBH [EC 3.2.1.91]) (Bayer et al. 1998), and two β-glucosidases (BGL [EC 3.2.1.21]) (Sang-Mok and Koo 2001). Endoglucanases act randomly against the amorphous region of the cellulose chain to produce reducing and nonreducing ends for cellobiohydrolases, which produce cellobiose from reducing or nonreducing ends of crystalline cellulose. Cellulose chains are thus efficiently degraded to soluble cellobiose and cellooligosaccharides by the endo-exo synergism of EG and CBH (Hebeish and Ibrahim 2007). In the last step of enzymatic cellulose degradation, cellooligosaccharides are hydrolyzed to glucose by β-glucosidase. In addition to endo-exo synergism, exo-exo synergism between two cellobiohydrolases has also been reported.

# Fungal Hemicellulases

Several different enzymes are needed to hydrolyze hemicelluloses, due to their heterogeneity (Saha2003). Xylan is the most abundant component of hemicellulose contributing over 70% of its structure. Xylanases are able to hydrolyze β-1, 4 linkages in xylan and produce oligomers which can be further hydrolyzed into xylose by β-xylosidase. Not surprisingly, additional enzymes such as β-mannanases, arabinofuranosidases or α-L-arabinases are needed depending on the hemicellulose composition which can be mannan-based or arabinofuranosyl-containing. Also similarly to cellulases, most of the hemicellulases are glycoside hydrolases (GHs), although some hemicellulases belong to carbohydrate esterases (CEs) which hydrolyze ester linkages of acetate or ferulic acid side groups (Shallom and Shoham 2003). A mixture of hemicellulases or pectinases with cellulases exhibited a significant increase in the extent of cellulose conversion (Ghose and Bisaria 1979, Beldman et al.1984). Many fungal species such as *Trichoderma*, *Penicillium*, *Aspergillus* and *T. emersonii* have been reported to produce large amounts of extracellular cellulases and hemicellulases.

## Fungal Ligninases

Fungi degrade lignin by secreting enzymes collectively termed "ligninases". These include two ligninolytic families; i) phenol oxidase (laccase) and ii) peroxidases [lignin peroxidase (LiP) and manganese peroxidase (MnP)] (Martinez et al. 2005). White-rot basidiomycetes such as *Coriolus versicolor* (Wang et al. 2003), *P. chrysosporium* and *T. versicolor* (Moredo et al. 2003) have been found to be the most efficient lignin-degrading microorganisms studied. Interestingly, LiP is able to oxidize the non-phenolic part of lignin, but it was not detected in many lignin degrading fungi. In addition, it has been widely accepted that the oxidative ligninolytic enzymes are not able to penetrate the cell walls due to their size. Thus, it has been suggested that prior to the enzymatic attack, low-molecular weight diffusible reactive oxidative compounds have to initiate changes to the lignin structure and hemicellulose, fungal cellulosomes are much less well characterized compared to bacterial cellulosomes.

## Fermentation

In the fermentation process, the hydrolytic products including monomeric hexoses (glucose, mannose and galactose) and pentoses (xylose and arabinose) will be fermented to valuable products such as ethanol. Among these hydrolytic products, glucose is normally the most abundant, followed by xylose or mannose and other lower concentration sugars.

The last two steps of bioconversion of pretreated lignocellulolytic residues to ethanol (hydrolysis and fermentation) can be performed separately (SHF) or simultaneously (SSF). In the separate hydrolysis and fermentation (SHF), the hydrolysate products will be fermented to ethanol in a separate process. The advantage of this method is that both processes can be optimized individually (e.g. optimal temperature is 45-50°C for hydrolysis, whereas it is 30°C for fermentation). However, its main drawback is the accumulation of enzyme-inhibiting end-products (cellobiose and glucose) during the hydrolysis. This makes the process inefficient, and the costly addition of β-glucosidase is needed to overcome end-product inhibition (Elumalia and Thangavelu 2010).

Further process integration can be achieved by a process known as consolidated bioprocessing (CBP) which aims to minimize all bioconversion steps into one step in a single reactor using one or more microorganisms. CBP operation featuring cellulase production, cellulose/hemicellulose hydrolysis and fermentation of 5- and 6- carbon sugars in one step have shown the potential to provide the lowest cost for biological conversion of cellulosic biomass to fuels, when processes relying on hydrolysis by enzymes and/or microorganisms are used (Lynd et al. 2005).

The simultaneous saccharification and fermentation (SSF) process was first studied by Takagi et al. (1977) for cellulose conversion to ethanol. The SSF process was originally developed for lignocellulosic biomass by researchers at Gulf Oil Company in 1974 (Blotkamo et al. 1978). The SSF process eliminates expensive equipment and reduces the probability of contamination by unwanted organisms that are less ethanol tolerant than the microbes selected for fermentation (Szczodrak 1989).

SSF combines the enzymatic saccharification of polymeric cellulose to simple monomeric forms such as glucose and its eventual fermentation by yeast to ethanol in the same vessel (Ikwebe and Harvey2011). In simultaneous saccharification and fermentation (SSF), however, the end-products will be directly converted to ethanol by the microorganism. Therefore, addition of high amounts of β-glucosidase is not necessary and this reduces the ethanol production costs (Stenberg et al. 2000). Rapid conversion of the glucose into ethanol by yeast results in faster rates, higher yields, and greater ethanol concentrations than possible for SHF. The presence of ethanol in the fermentation broth also makes the mixture less vulnerable to invasion by unwanted microorganisms (Sasikumar and Viruthagiri 2010). However, the main drawback of SSF is the need to compromise processing conditions such that temperature and pH are suboptimal for each individual step. The development of recombinant yeast strains with improved thermotolerance can enhance the performance of SSF (Galbe and Zacchi 2002). It is reported that the major inefficiencies of biochemical process for lignocellulosic bioethanol production were identified as the simultaneous saccharification and fermentation (SSF) process accounting for 27% of the lost energy by thermodynamic analysis (Sohel and Jack 2010). Alkasrawi et al. ( 2003) reported that addition of surfactants as an additive in SSF can significantly lower the operational cost of the process because it increases the conversion rate of cellulose to glucose. Addition of Tween-20, 2.5 g/l not only reduces the time required to attain maximum ethanol concentration, but also enhances enzyme activity in the liquid fraction at the end of SSF, probably by preventing unproductive binding of the cellulases to lignin, which could facilitate enzyme recovery.

Over the years, various groups have worked on the SSF process to improve the choice of enzymes, fermentative microbes, biomass pretreatment, and process conditions. Extensive studies on SSF have since been conducted focusing on the production of ethanol from cellulosic substrates. Phillipidis et al. (1993) have studied the enzymic hydrolysis of cellulose in an attempt to optimize SSF performance. Ghose et al. (1984) have increased ethanol productivity by employing a vacuum cycling in an SSF process using lignocellulosic substances. Zhu et al. (2006b) evaluated the suitability of production of ethanol from the microwave-assisted alkali pretreated wheat straw, the simultaneous saccharification and fermentation (SSF) of the microwave-assisted and conventional alkali pretreated wheat straw to ethanol.

*Candida brassicae* is accepted as the yeast of choice as far as SSF is considered, although both*Saccharomyces cerevisiae* and *S. carlsbergensis* have been found to offer similar rates. Several other yeasts as well as the bacteria *Zymomonas mobilis* have been studied with cellulose from *T. ressei*mutants for SSF processes. Researchers have also examined several combinations of enzymes with *Z.mobilis, S. cerevisiae*, and other ethanol producer, but they have only considered substrate levels lower than necessary to prove economic viability. Wyman et al. (1986) evaluated the cellobiose-fermenting yeast *Brettanomyces clausenii* for the SSF of cellulose to ethanol.

There are number of different methods to quantitate ethanol in samples. HPLC has been utilized to monitor the fermentation process This method has the advantage of being able to monitor not only the production of ethanol, but also the reaction substrates and byproducts (Hall and Reuter 2007). Fourier transform infrared spectroscopy (Sharma et al. 2009), gas chromatography (Wang et al.2003), and Infrared (Lachenmeier et al. 2010) technologies have also been used to detect and quantitate ethanol in samples. While FTIR requires a large investment in instrumentation, the use or less expensive IR technology has been demonstrated to be just as accurate (Lachenmeier et al.2010). Gerchman et al. (2012) developed a cheap and rapid approach for ethanol quantification in aqueous media during fermentation steps as part of the conversion of biomass to ethanol. The suggested method requires a sample of a small volume and consists of organic extraction, followed by direct use of gas chromatography with a flame ionization detector (GC-FID). The feasibility of such approach is obvious since there is no need for the head-space system, distillation, expensive reagents and sophisticated equipment. The proposed method was also tested for its 'real-life' applicability for ethanol quantification from fermentation process.

According to Keller and Bryan (2000), distillation is still a "formidable competitor" as a major separation method even though much research has been thrust on its alternatives. Hence, distillation, especially simple distillation, tends to be the first choice in industry for separating a liquid mixture; other methods, including complex distillation, e.g., azeotropic distillation, come into play only when simple distillation is deemed to be technically infeasible or economical inviable because of typically three large stainless steel distillation towers, stainless steel heat exchangers and price of stainless up 400% in last six years, high operating costs because 280 MMBTU/hr energy is consumed (100 MGPY ethanol). Mole sieve drying adds to energy costs and that's why energy costs up significantly with price of crude oil.

Under certain circumstances, retrofitting of an existing process can be economically far more viable than constructing a new process, especially when the financial resources are limited and/or when short term needs are to be met under a tight time constraint. Developing economically viable fermentation processes requires efficient downstream processing: selective product removal and avoiding byproduct streams. "ESepis a modular, low-energy process for the recovery of ethanol from fermentation broth with an estimated reduction of up to 60% in both capital and operating costs versus conventional distillation. Use of non-stainless steel components also results in a substantial reduction in construction time". It is applicable to new ethanol plants (corn, sugar and cellulosic). It replaces whole distillation train and mole sieve dryer. With new plants it reduces overall energy consumption by >60% (ESep 2008).

The utilization of pervaporation for the production of absolute (anhydrous) ethanol through its coupling with the previous distillation step has been reported. The modeling and optimization of the process using MINLP tools showed 12% savings in the production costs considering a 32% increase in membrane area and the reduction in both reflux ratio and ethanol concentration in the distillate of the column (Lelkes et al. 2000, Szitkai et al. 2002). Through pilot-plant studies, the integration of distillation process with the pervaporation has been achieved resulting in good indexes in terms of energy savings. These savings are due to the low operation costs of pervaporation and to the high yield of dehydrated ethanol, typical of pervaporation processes (Tsuyomoto et al. 1997). The comparison between azeotropic distillation using benzene and pervaporation system using multiple membrane modules showed that, at the same ethanol production rate and quality (99.8 wt.%), operation costs, including the membrane replacement every 2–4 years, are approximately 1/3–1/4 of those of azeotropic distillation.

# Methods Used to Improve Fungal Enzyme Production, Activity and/or Stability

In order to produce ethanol industrially, the fermentative microorganism needs to be robust. The utilization of all the sugars generated from lignocellulosic hydrolysate is essential for the economical production of ethanol (Saha 2003). The conventional ethanol fermenting yeast (Saccharomyces cerevisiae) or bacterium (Zymomonas mobilis) cannot ferment multiple sugar substrates to ethanol (Bothast et al. 1999). A major technical hurdle to converting lignocellulose to ethanol is developing an appropriate microorganism for the fermentation of a mixture of sugars such as glucose, xylose, arabinose, and galactose (Bothast et al. 1999). A number of recombinant microorganisms such as*Escherichia coli, Klebsiella oxytoca, Z. mobilis,* and *S. cerevisiae* have been developed over the last 25 years with a goal of fermenting mixed sugars to ethanol (Alterthum and Ingram 1989, Ohta et al.1991, Zhang et al. 1995, Ho et al. 1998). Saha and Cotta's (2011) research unit has developed a recombinant *E. coli* (strain FBR5) that can ferment mixed multiple sugars to ethanol (Dien et al.2000). The strain carries the plasmid pLOI297, which contains the genes for pyruvate decarboxylase (pdc) and alcohol dehydrogenase (adh) from *Z. mobilis* necessary for efficiently converting pyruvate into ethanol (Alterthum and Ingram 1989).

Technologies required for bioconversion of lignocelluloses to ethanol and other valuable products are currently available but need to be developed further in order to make biofuels cost competitive compared to other available energy resources such as fossil fuels. The most recent and important improvements in production/activity of fungal enzymes using different techniques such as mutagenesis, co-culturing and heterologous gene expression of cellulases are discussed below.

# Mutagenesis

Many fungal strains have been subjected to extensive mutagenesis studies due to their ability to secrete large amounts of cellulose-degrading enzymes. Cellulolytic activity of *T. reesei* QM6a has been improved by using different mutagenesis techniques including UV-light and chemicals, resulting in the mutant QM 9414 with higher filter paper activity (FPA) (Mandels et al. 1971). *T. reesei* RUT-C30 is one of the best known mutants, producing 4–5 times more cellulase than the wild-type strain (QM 6a). A recent study by Kovacs et al. 2008 has shown that wild-type *Trichoderma atroviride* (F-1505) produces the most cellulase among 150 wild-type *Trichoderma*. Moreover, *T. atroviride*mutants were created by mutagenesis using *N*-methyl-*N'*-nitro-*N*-nitrosoguanidine (NTG) as well as UV-light. These *T. atroviride* mutants (e.g. *T. atroviride* TUB F-1724) produce high levels of extracellular cellulases as well as β-glucosidase when they are grown on pretreated willow. Cellulase and xylanase activities in *Penicillium verruculosum* 28 K mutants were improved about 3-fold using four cycles of UV mutagenesis. The enzyme production was further improved by 2- to 3-fold in a two-stage fermentation process using wheat bran, yeast extract medium and microcrystalline cellulose as the inducer (Soloveva et al. 2005).

Site-directed mutagenesis (SDM) plays a central role in the characterization and improvement of cellulases including their putative catalytic and binding residues. The application of SDM it was found that Glu 116 and 200 are the catalytic nucleophile and acid–base residues in *Hypocrea jecorina*(anamorph *T. reesei*) Cel12A, respectively. In the study, mutant enzymes were produced where Glu was replaced by Asp or Gln at each position (E116D/Q and E200D/Q). The specific activity of these mutants was reduced by more than 98%, suggesting the critical role of these two residues in the catalytic function of the enzyme (Okada et al. 2000). In another study, the thermostable endo-1, 4-β-xylanase (XynII) mutants from *T. reesei* were further mutated to resist inactivation at high pH by using SDM. All mutants were resistant to thermal inactivation at alkaline pH. For example, thermotolerance for one mutant (P9) at pH 9 was increased approximately 4–5°C, resulting in better activity in sulphate pulp bleaching compared to the reference (Fenel et al. 2006). Also, the catalytic efficiency and optimum pH of *T. reesei* endo-β-1, 4-glucanase II were improved by saturation mutagenesis followed by random mutagenesis and two rounds of DNA shuffling. The pH optimum of the variant (Q139R/L218H/W276R/N342T) was shifted from 4.8 to 6.2, while the enzyme activity was improved more than 4.5-fold (Qin et al. 2008). Moreover, the stability of *T. reesei* endo-1, 4-β-xylanases II (XynII) was increased by engineering a disulfide bridge at its N-terminal region. In fact, two amino acids (Thr-2 and Thr-28) in the enzyme were substituted by cysteine (T2C:T28C mutant) resulting in a 15°C increase in thermostability (Fenel et al. 2004).

# Co-Culturing

Fungal co-culturing offers a means to improve hydrolysis of lignocellulosic residues, and also enhances product utilization which minimizes the need for additional enzymes in the bioconversion process. In the case of cellulose degradation, for example, all three enzymatic components (EG, CBH and β-glucosidase) have to be present in large amounts. However, none of the fungal strains, including the best mutants, are able to produce high levels of the enzymes at the same time. *T.reesei* for example produces CBH and EG in high quantities whereas its β-glucosidase activity is low (Stockton et al. 1991). *A. niger* however, produces large amounts of β-glucosidase, but has limited EG components (Kumar et al. 2008). In addition, hemicellulose hydrolysis must also be considered when lignocellulosic residues are subjected to biomass conversion. However, this will be determined by the pretreatment methods. Specifically in an alkali pretreatment method, a part of lignin will be removed and thus hemicellulose has to be degraded by the use of hemicellulases, whereas in acid-catalyzed pretreatment, the hemicellulose layer will be hydrolyzed (Hahn et al. 2006). Again, some fungal strains have been shown to work more efficiently on cellulosic residues whereas others produce more hemicellulolytic enzymes and efficiently hydrolyze hemicellulosic portions (Howard et al. 2003). Conversion of both cellulosic and hemicellulosic hydrolytic products in a single process can be achieved by co-culturing two or more compatible microorganisms with the ability to utilize the materials. In fact, in nature, lignocellulosic residues are degraded by multiple co-existing lignocellulolytic microorganisms.

Mixed fungal cultures have many advantages compared to their monocultures, including improving productivity, adaptability and substrate utilization. Improving fungal cellulolytic activity of *T. reesei* and *A. niger* by co-culturing was the subject of extensive research including studies done by Maheshwari et al. (1994), Ahamed et al. (2008) and Juhasz et al. 2003. Moreover, other fungal strains have been co-cultured to obtain better cellulolytic activity such as co-culturing of *T. reesei*RUT-C30 and *A. phoenicis* (Duff 1985) or *A. ellipticus* and *A. fumigatus* (Gupte and Madamwar1997). There are a few examples of co-culturing fungal strains for the purpose of combining cellulose and hemicellulose hydrolysis such as co-culturing *T. reesei* D1-6 and *A. wentii* Pt 2804 in a mixed submerged culture (Panda et al. 1983) or co-culturing *T. reesei* LM-UC4 and *A. phoenicis* QM329 using ammonia-treated bagasse (Duenas et al. 1995). In the both cases, enzyme activity for cellulases and hemicellulases was significantly increased. The main drawback of co-culturing however is the complexity of growing multiple microorganisms in the same culture (Lynd et al. 2002).

# Metabolic Engineering

Metabolic engineering is a powerful method to improve, redirect, or generate new metabolic reactions or whole pathways in microorganisms. This enables one microorganism to complete an entire task from beginning to end. This can be done by altering metabolic flux by blocking undesirable pathway(s) and/or enhancement of desirable pathway(s). For example by application of homologous recombination, the production of *T. reesei* β-glucosidase I was enhanced using xylanase (*xyn3*) and cellulase (*egl3*) promoters which improved β-glucosidase activity to 4.0 and 7.5 fold compared to the parent, respectively. This will permit one fungal strain such as *T. reesei* to be more efficient on hydrolysis of cellulose to glucose which improve the yield and therefore lower the cost (Rahman et al.2009). Becker and Boles (2003) described the engineering of a *Saccharomyces cerevisiae* strain able to utilize the pentose sugar L-arabinose for growth and to ferment it to ethanol. Expanding the substrate fermentation range of *S. cerevisiae* to include pentoses is important for the utilization of this yeast in economically feasible biomass-to-ethanol fermentation processes. After overexpression of a bacterial L-arabinose utilization pathway consisting of *Bacillus subtilis* AraA and *Escherichia coli* AraB and AraD and simultaneous overexpression of the L-arabinose-transporting yeast galactose permease, we were able to select an L-arabinose-utilizing yeast strain by sequential transfer in L-arabinose media. High L-arabinose uptake rates and enhanced transaldolase activities favor utilization of L-arabinose.

Shaw et al. 2008 engineered *Thermoanaerobacterium saccharolyticum*, a thermophilic anaerobic bacterium that ferments xylan and biomass-derived sugars, to produce ethanol at high yield. Knockout of genes involved in organic acid formation (acetate kinase, phosphate acetyltransferase, and L-lactate dehydrogenase) resulted in a strain able to produce ethanol as the only detectable organic product and substantial changes in electron flow relative to the wild type. Glucose and xylose are co-utilized and utilization of mannose and arabinose commences before glucose and xylose are exhausted.

# Heterologous Expression

Heterologous expression is a powerful technique to improve production yield of enzymes, as well as activity. In order to make a robust lignocellulolytic fungal strain, many different fungal cellulases with higher and/or specific activity based on the need for a functional cellulase system in the organism have been cloned and expressed. For example, thermostable β-glucosidase (cel3a) from thermophilic fungus *T. emersonii* was expressed in *T. reesei* RUT-C30 using a strong *T. reesei cbh1* promoter. The expressed enzyme has been shown to be highly thermostable (optimum temperature at 71.5°C) with high specific activity (Murray et al. 2004). In the study for the improvement of biofinishing of cotton,*T. reesei* cellobiohydrolase (I & II) were overexpressed using additional copy(s) of the genes cloned under *T. reesei cbh1* promoter. The results have shown that the expression of CBHI was increased to 1.3- and 1.5-fold with one or two additional copies of the gene, respectively.

# Immobilization

Immobilization of microbial cells and enzymes have showed certain technical and economical advantages over free cell system. Using immobilized enzymes not only leads to greater product purity, cleaner processes, and economic operational costs but also makes the use enzyme cost effective and recoverable (Meena and Raja 2004). The immobilized biocatalysts have been extensively investigated during last few decades. An immobilized cellobiase enzyme system has been used in the enzymatic hydrolysis of biomass for the generation of cellulosic ethanol (Das et al., 2011). Production of alcohol and biodiesel fuel from triglycerides using immobilized lipase has been carried out using porous kaolinite particle as a carrier (Iso et al. 2001).

The use of an immobilized yeast cell system for alcoholic fermentation is an attractive and rapidly expanding research area because of its additional technical and economic advantages compared with the free cell system. A reduction in the ethanol concentration in the immediate microenvironment of the organism due to the formation of a protective layer or specific adsorption of ethanol by the support may act to minimize end product inhibition. The most significant advantages of immobilized yeast cell systems are the ability to operate with high productivity at dilution rates exceeding the maximum specific growth rate, the increase of ethanol yield and cellular stability and the decrease of process expenses due to the cell recovery and reutilization (Lin and Tanaka 2006). Other advantages of immobilized cell system over presently accepted batch or continuous fermentations with free-cells are: greater volumetric productivity as a result of higher cell density; tolerance to higher concentrations of substrate and products; lacking of inhibition; relative easiness of downstream processing etc. in different types of bioreactors, such as packed bed reactor, fluidized bed reactor, gaslift reactor and reactor with magnetic field (Ivanova et al. 1996, Sakai et al. 1994; Perez et al.2007).

Perspective techniques for yeasts immobilization can be divided into four categories: attachment or adsorption to solid surfaces (wood chips, delignified brewer's spent grains, DEAE cellulose, and porous glass), entrapment within a porous matrix (calcium alginate, k-carrageenan, polyvinyl alcohol, agar, gelatine, chitosan, and polyacrilamide), mechanical retention behind a barrier (microporous membrane filters, and microcapsules) and self-aggregation of the cells by flocculation (Ivanova et al.2011).

# Process Integration

One of the most important approaches for the design of more intensive and cost-effective process configurations is process integration. Process integration looks for the integration of all operations involved in the production of fuel ethanol. This can be achieved through the development of integrated bioprocesses that combine different steps into one single unit. Thus, reaction–separation integration by removing ethanol from the zone where the biotransformation takes place, offers several opportunities for increasing product yield and consequently reducing product costs. Other forms of integration may significantly decrease energetic costs of specific flow sheet configurations for ethanol production. Process integration is gaining more and more interest due to the advantages related to its application in the case of ethanol production: reduction of energy costs, decrease in the size and number of process units, intensification of the biological and downstream processes. Integration of fermentation and separation processes for reduction of product inhibition, development of efficient cogeneration technologies using cane bagasse, development of CBP, application of membrane technology (e.g. for ethanol removal or dehydration) are examples of process integration.

# CONCLUSIONS

Lignocellulolytic microorganisms, especially fungi, have attracted a great deal of interest as biomass degraders for large-scale applications due to their ability to produce large amounts of extracellular lignocellulolytic enzymes. Many successful attempts have been made to improve fungal lignocellulolytic activity including recombinant and non-recombinant techniques. Process integration has also been considered for the purpose of decreasing the production cost, which was partly achieved by performing hydrolysis and fermentation in a single reactor (SSF) using one or more microorganisms (co-culturing).

These laboratory improvements should now be verified in pilot and demonstration plants. Scaling up the production of lignocellulosic ethanol, however, requires further reduction of the production cost. Thus, in order to improve the technology and reduce the production cost, two major issues have to be addressed: i) improving technologies to overcome the recalcitrance of cellulosic biomass conversion (pretreatment, hydrolysis and fermentation) and ii) sustainable production of biomass in very large amounts.

## Future Prospects

It is considered that lignocellulosic waste will become the main feedstock for ethanol production in the near future. In the case of large scale biomass production, additional waste stocks can be tested and used as substrates to meet the needs. On the other hand, biotechnological approaches including systems biology and computational tools are likely good candidates to overcome these issues. Future trends for costs reduction should include more efficient pretreatment of biomass, improvement of specific activity and productivity of cellulases, improvement of recombinant microorganisms for a greater assimilation of all the sugars released during the pretreatment and hydrolysis processes, and further development of co-generation system. Undoubtedly, ongoing research on genetic and metabolic engineering will make possible the development of effective and stable strains of microorganisms for converting cellulosic biomass into ethanol. Process engineering will play a central role for the generation, design, analysis and implementation of technologies improving the indexes of global process, or for the retrofitting of employed bioprocesses. Undoubtedly, process intensification through integration of different phenomena and unit operations as well as the implementation of consolidated bioprocessing of different feedstock's into ethanol (that requires the development of tailored recombinant microorganisms), will offer the most significant outcomes during the search of the efficiency in fuel ethanol production. This fact will surely imply a qualitative improvement in the industrial production of fuel ethanol in the future.

# ACKNOWLEDGEMENTS

The authors are thankful to Chancellor, ITM University, Gwalior, MP, and Dr. J.L. Bhat, Dean, Life Sciences, ITM University, Gwalior, MP, for their constant support and encouragement.

# REFERENCES

1.    Ahamed A, Vermette P (2008) Enhanced enzyme production from mixed cultures of Trichoderma reesei RUT-C30 and Aspergillus niger LMA grown as fed batch in a stirred tank bioreactor. Biochem Eng J 42:41-46

2.    Alinia R, Zabihi S, Esmaeilzadeh F, Kalajahi JF (2010) Pretreatment of wheat straw by supercritical $CO_2$ and its enzymatic hydrolysis for sugar production. Biosystems Eng 107(1):61-66

3.    Alizadeh H, Teymouri F, Gilbert TI, Dale BE (2005) Pretreatment of switchgrass by ammonia fiber explosion (AFEX). Appl Biochem Biotechnol 124(1–3):1133-1141

4.    Alkasrawi M, Eriksson T, Borjesson J, Wingren A, Galbe M, Tjerneld F, Zacchi G (2003) The effect of tween-20 on simultaneous saccharification and fermentation of softwood to ethanol. Enzyme and Microbial Technol 33:71-78

5.    Alterthum F, Ingram LO (1989) efficient ethanol production from glucose, lactose, and xylose by recombinant Escherichia coli. Appl Environ Microbiol 55:1943-1948.

6.    Anderson WF, Akin DE (2008) Structural and chemical properties of grass lignocelluloses related to conversion for biofuels. J Ind Microbiol Biotechnol 35:355-366

7.    Arora DS, Chander MKGP (2002) Involvement of lignin peroxidase, manganese peroxidase and laccase in degradation and selective ligninolysis of wheat straw. Int Bioterior Biodegrad 50:115-120

8.    Aswathy US, Sukumaran RK, Lalitha D, Rajeshree KP, Singhania RR, Pandey A (2010) Bio-ethanol from water hyacinth biomass: an evaluation of enzymatic saccharification strategy. Bioresour Technol 101:925-930

9.    Azzam M (1989) Pretreatment of cane bagasse with alkaline hydrogen peroxide for enzymatic hydrolysis of cellulose and ethanol fermentation. J Environ Sci Health B 24(4):421-433

10.   Balan V, Sousa LDC, Chundawat SPS, Marshall D, Sharma LN, Chambliss CK, Dale BE (2009) Enzymatic digestibility and pretreatment degradation products of AFEX-treated hardwoods (Populusnigra). Biotechnol Prog 25:365-375

11.   Ballesteros I, Negro MAJ, Oliva JM, Cabanas A, Manzanares P, Ballesteros M (2006) Ethanol production from steam-explosion pretreated wheat straw. Appl Biochem Biotechnol 130(1–3):496-508

12.   Ballesteros M, Oliva JM, Negro MJ, Manzanares P, Ballesteros I (2004) Ethanol from lignocellulosic materials by a simultaneous saccharification and fermentation process (SFS) with Kluyveromyces marxianus CECT 10875. Process Biochem 39:1843-1848

13.   Banerjee S, Sen R, Pandey RA (2009) Evaluation of wet air oxidation as a pretreatment strategy for bioethanol production from rice husk and process optimization. Biomass Bioenergy 33(12):1680-1686

14.   Bayer EA, Chanzy H, Lamed R, Shoham Y (1998) Cellulose, cellulases and cellulosomes. Curr Opin Struct Biol 8(5):548-557

15. Becker J, Boles E (2003) A modified Saccharomyces cerevisiae strain that consumes l-arabinose and produces ethanol. Appl Environ Microbiol 69(7):4144-4150

16. Beldman G, Rombouts FM, Voragen AGJ, Pilnik W (1984) Application of cellulase and pectinase from fungal origin for the liquefaction and saccharification of biomass. Enzyme Microb Technol 6:503-507

17. Betancur GJV (2005) Avanços em biotecnologia de hemicelulose Para produçao de etanol por Pichia stipitis. Dissertaçao de Mestrado, Escola de Quimica da UFRJ, Rio de Janeiro.

18. Bhatia L, Paliwal S (2010) Banana peel waste as substrate for ethanol production. International J of Biotechnol and Bioeng Research 1(2):213-218

19. Bhatia L, Paliwal S (2011) Ethanol producing potential of Pachysolen tannophilus from sugarcane bagasse. International J of Biotechnol and Bioeng Research 2(2):271-276

20. Bisaria VS (1998) Bioprocessing of agro-residues to value added products. In: Martin AM (ed) Bioconversion of waste materials to industrial products, 2nd edn. Chapman and Hall, UK. pp 197-246

21. Bjerre AB, Olesen AB, Fernqvist T (1996) Pretreatment of wheat straw using combined wet oxidation and alkaline hydrolysis resulting in convertible cellulose and hemicellulose. Biotechnol Bioeng 49:568-577

22. Blotkamo PJ, Takaai M, Pemberton MS, Enert GS (1978) Enzymatic hydrolysis of cellulose and simultaneous saccharification to alcohol. AIChE Symp Ser 74:85

23. Bothast RJ, Nichols NN, Dien BS (1999) Fermentation with new recombinant organisms. Biotechnol Prog 15:867-875

24. Brooks AA (2008) Ethanol production potential of local yeast strains isolated from ripe banana peels. African J of Biotechnol 7(20):3749-3752

25. Cadoche L, Lopez GD (1989) Assessment of size reduction as a preliminary step in the production of ethanol from lignocellulosic wastes. Biol Wastes 30:153-157

26. Cao Y, Li H, Zhang Y, Zhang J, He J (2010) Structure and properties of novel regenerated cellulose films prepared from cornhusk

cellulose in room temperature ionic liquids. J Appl Polymer Sci 116(1):547-554

27.    Champagne P (2007) Feasibility of producing bio-ethanol from waste residues: a Canadian perspective feasibility of producing bio-ethanol from waste residues in Canada. Resour Conserv Recycl 50:211-230

28.    Cheng YS, Zheng Y, Yu CW, Dooley TM, Jenkins BM, Vander Gheynst JS (2010) Evaluation of high solids alkaline pretreatment of rice straw. Appl Biochem Biotechnol 162(6):1768-1784

29.    Chornet E, Vanasse C, Lemonnier JP, Overend RP (1988) Preparation and processing of medium and high consistency biomass suspensions. Research in thermochemical biomass conversion. 766-778

30.    Chundawat SPS, Venkatesh B, Dale BE (2007) Effect of particle size based separation of milled corn stover on AFEX pretreatment and enzymatic digestibility. Biotechnol Bioeng 96(2):219-231

31.    Coveca (2002) Comision veracruzana de comercializacion agropecuaria. Gobierno del Estado de Veracruz, Mexico.

32.    Curreli N, Agelli M, Pisu B, Rescigno A, Sanjust E, Rinaldi A (2002) Complete and efficient enzymic hydrolysis of pretreated wheat straw. Process Biochem 37:937-941

33.    Curreli N, Fadda MB, Rescigno A, Rinaldi AC, Soddu G, Sollai F (1997) Mild alkaline/oxidative pretreatment of wheat straw. Process Biochem 32:665-670

34.    Dadi AP, Schall CA, Varanasi S (2007) Mitigation of cellulose recalcitrance to enzymatic hydrolysis by ionic liquid pretreatment. Appl Biochem Biotechnol 12:407-421

35.    Das S, Schlessel DB, Ji HF, Donough JM, Wei Y (2011) Enzymatic hydrolysis of biomass with recyclable use of cellobiase enzyme immobilized in sol–gel routed mesoporous silica. J Mol Catal B Enzym 70:49-54

36.    Dashtban M, Schraft H, Qin W (2009) Fungal bioconversion of lignocellulosic residues, opportunities & perspectives. Int J Biol Sci 5(6):578-595

37.    Dawson L, Boopathy R (2007) Use of post-harvest sugarcane residue for ethanol production. Biores Technol 98:1695-1699

38. De Moraes-Rocha GJ, Martin C, Soares IB, Maior AMS, Baudel HM, De Abreu CAM (2010) Dilute mixed-acid pretreatment of sugarcane bagasse for ethanol production. Biomass Bioenergy 35:663-670

39. Delmer DP, Amor Y (1995) Cellulose biosynthesis. Plant Cell 7:987-1000

40. Demain AL, Michael Newcomb WJHD (2005) Cellulase, clostridia, and ethanol. Microbiol Mol Biol Rev 69(1):124-154

41. Dhillon GS, Bansal S, Oberoi HS (2007) Cauliflower waste incorporation into cane molasses improves ethanol production using Saccharomyces cerevisiae MTCC 178. Indian J Microbiol 47:353-357

42. Dien BS, Nichols NN, O'Bryan PJ, Bothast RJ (2000) Development of new ethanologenic Escherichia coli strains for fermentation of lignocellulosic biomass. Appl Biochem Biotechnol 84–86:181-186

43. Dienes D, Egyhazi A, Reczey K (2004) Treatment of recycled fiber with trichoderma cellulases. Ind Crop Prod 20(1):11-21

44. Duenas R, Tengerdy RP, Gutierrez-Correa M (1995) Cellulase production by mixed fungi in solid-substrate fermentation of bagasse. World J of Microbiol & Botechnol 11:333-337

45. Duff SJB, Murray WD (1996) Bioconversion of forest products industry waste cellulosics to fuel ethanol: a review. Bioresour Technol 55:1-33

46. Duff SJB (1985) Cellulase and beta-glucosidase production by mixed culture of Trichoderma reesei Rut C30 and Aspergillus phoenicis. Biotechnol Lett 7:185-190

47. Ebringerova A, Hromadkova Z (2002) Effect of ultrasound on the extractibility of corn bran hemicelluloses. Ultrason Sonochem 9(4):225-229

48. ElSeoud OA, Koschella A, Fidale LC, Dorn S, Heinze T (2007) Applications of ionic liquids in carbohydrate chemistry: a window of opportunities. Biomacromolecules 8(9):2629-2647

49. Elumalia S, Thangavelu V (2010) Simultaneous saccharification and fermentation (SSF) of pretreated sugarcane bagasse using cellulose and Saccharomyces cerevisiae-kinetics and modeling. Chem Eng Res Bull 14:29-35

50.  ESep (2008) A Novel Low Energy Route to Ethanol Recovery, Trans Ionics Corporation21st NREL Industry Growth Forum October 28-30, Trans Ionics Corporation. Incorporated 2000. Energy-Saving Separation.

51.  Fan LT, Gharpuray MM, Lee YH (1987) Cellulose hydrolysis biotechnology monographs. Springer, Berlin. p 57

52.  Fenel F, Zitting AJ, Kantelinen A (2006) Increased alkali stability in Trichoderma reesei endo-1, 4-beta-xylanase II by site directed mutagenesis. J Biotechnol 121:102-107

53.  Fenel F, Leisola M, Janis J, Turunen O (2004) A de novo designed N-terminal disulphide bridge stabilizes the trichoderma reesei endo-1,4-beta-xylanase II. J Biotechnol 108:137-143

54.  Ferraz A, Ana M, Cordova A (2003) Machuca: wood biodegradation and enzyme production by Ceriporiopsis subvermispora during solid-state fermentation of Eucalyptus grandis. Enzyme and Microbial Technol 32:59-65

55.  Focher B, Marzett A, Crescenzi V (eds) (1991) Steam explosion techniques, fundamentals and industrial applications Gordon and Breach, Philadelphia, Pa, USA.

56.  Galbe M, Zacchi G (2002) A review of the production of ethanol from softwood. Appl Microbiol Biotechnol 59:618-628

57.  Gama FM, Carvalho MG, Figueiredo MM, Mota M (1997) Comparative study of cellulose fragmentation by enzymes and ultrasound. Enzyme Microb Technol 20:12-17

58.  Gerchman Y, Schnitzer A, Gal R, Mirsky N, Chinkov N (2012) A simple rapid gas-chromatography flame-ionization-detector (GC-FID) method for the determination of ethanol from fermentation processes African J of. Biotechnol 11(15):3612-3616

59.  Ghose TK, Bisaria VS (1979) Studies on mechanism of enzymatic hydrolysis of cellulosic substances. Biotechnol Bioeng 21:131-146

60.  Ghose TK, Roychoudhury PK, Ghosh P (1984) Simultaneous saccharification and fermentation (SSF) of lignocellulosics to ethanol under vacuum cycling and step feeding. Biotechnol Bioeng 26:377-381

61.  Goldstein I, Easter J (1992) An improved process for converting cellulose to ethanol. Tappi J 28:135

62. Grassick A, Murray PG, Thompson R, Collins CM, Byrnes L, Birrane G, Higgins TM, Tuohy MG (2004) Three-dimensional structure of a thermostable native cellobiohydrolase, CBH IB, and molecular characterization of the cel7 gene from the filamentous fungus, Talaromyces emersonii. Eur J Biochem 271:4495-4506

63. Gupte A, Madamwar D (1997) Solid state fermentation of lignocellulosic waste for cellulose and β-glucosidase production by co-cultivation by Aspergillus ellipticus and Aspergillus fumigatus. Biotechnol Prog 13:166-169

64. Ha MA, Apperley DC, Evans BW, Huxham IM, Jardine WG, Vietor RJ, Reis D, Vian B, Jarvis MC (1998) Fine structure in cellulose microfibrils: NMR evidence from onion and quince. Plant J 16:183-190

65. Hahn-Hagerdal B, Galbe M, Gorwa-Grauslund MF, Lidan G, Zacchi G (2006) Bio-ethanol-the fuel of tomorrow from the residues of today. Trends Biotechnol 24(12):549-556

66. Hall G, Reuter WM (2007) HPLC Analysis for the Monitoring of Fermentation Broth During Ethanol Production as a Biofuel Application brief, Perkin Elmer. http://www.perkinelmer.com/pdfs/downloads/ABR_EthanolAsBiofuelbyHPLCAppBrief.pdf

67. Hamelinck CN, Hooijdonk GV, Faaji APC (2005) Ethanol from lignocellulosic biomass: techno-economic performance in short-, middle- and long-term. Biomass Bioenergy 28:384-410

68. Hebeish A, Ibrahim NA (2007) the impact of frontier sciences on textile industry. Colourage 54:41-55

69. Himmel ME, Ding SY, Johnson DK, Adney WS, Nimlos MR, Brady JW, Foust TD (2007) Biomass recalcitrance: engineering plants and enzymes for biofuels production. Science 315(5813):804-807

70. Ho NWY, Chen Z, Brainard AP (1998) Genetically engineered saccharomyces yeast capable of effective cofermentation of glucose and xylose. Appl Environ Microbiol 64:1852-1859

71. Howard RL, Abotsi E, Jansen van Rensburg EL, Howard S (2003) Lignocellulose biotechnology: issues of bioconversion and enzyme production. African J of Biotechnol 2:602-619

72. Ibrahim MM, El-Zawawy WK, Abdel-Fattah YR, Soliman NA, Agblevor FA (2011) Comparison of alkaline pulping with steam

explosion for glucose production from rice straw. Carbohydr Polym 83(2):720-726

73. Ikweb J, Harvey AP (2011) Intensification of bioethanol production by simultaneous saccharification and fermentation (SSF) in an oscillatory baffled reactor (OBR). Bioenergy technol. World Renewable Energy Congress, Sweden. pp 381-388

74. Imai M, Ikari K, Suzuki I (2004) High-performance hydrolysis of cellulose using mixed cellulase species and ultrasonication pretreatment. Biochem Eng J 17:79-83

75. Isitua CC, Ibeh IN (2010) Novel method of wine production from banana (Musa acuminata) and pineapple (Ananas comosus) wastes. African J of Biotechnol 9(44):7521-7524

76. Iso M, Chen B, Eguchi M, Kudo T, Shrestha S (2001) Production of biodiesel fuel from triglycerides and alcohol using immobilized lipase. J Mol Catal B: Enzym 16(1):53-58

77. Ivanova V, Hristov J, Dobreva E, AlHassan Z, Penchev I (1996) Performance of a magnetically stabilized bed reactor with immobilized yeast cells. Appl Biochem Biotechnol 59:187-198

78. Ivanova V, Petrova P, Hristov J (2011) Application in the ethanol fermentation of immobilized yeast cells in matrix of alginate/magnetic nanoparticles, on chitosan-magnetite microparticles and cellulose-coated magnetic nanoparticles. Int Rev Chem Eng 3:289-299

79. Iyer PV, Wu ZW, Kim SB, Lee YY (1996) Ammonia recycled percolation process for pretreatment of herbaceous biomass. Appl Biochem Biotechnol 57–58:121-132

80. Jacobsen SE, Wyman CE (2000) Cellulose and hemicellulose hydrolysis models for application to current and novel pretreatment processes. Appl Biochem Biotechnol 84(1–9):81-96

81. Juhasz T, Kozma K, Szengyel Z, Reczey K (2003) Production of β-glucosidase in mixed culture of Aspergillus Niger BKMF 1305 and Trichoderma reesei RUT C30. Food Technol Biotechnol 41:49-53

82. Kalogo Y, Habibi S, MacLean HL, Joshi SV (2007) Environmental implications of municipal solid waste-derived ethanol. Environ Sci Technol 41:35-41

83. Kaur J, Chandha BS, Kumar BA (2007) Purification and characterization of β-glucosidase from melanocarpus Sp. MTCC3922. Elect J Biotechnol 10:260-270

84. Keller GE, Bryan PF (2000) Process engineering moving in New directions. Chem Eng Prog 96(1):41-50

85. Kilzer FJ, Broido A (1965) Speculations on the nature of cellulose pyrolysis. Pyrodynamics 2:151-163

86. Kim KH, Hong J (2001) Supercritical CO2 pretreatment of lignocellulose enhances enzymatic cellulose hydrolysis. Bioresour Technol 77(2):139-144

87. Kim S, Dale BE (2004) Global potential bioethanol production from wasted crops and crop residues. Biomass Bioenergy 26:361-375

88. Kim TH, Lee YY (2005) Pretreatment and fractionation of corn stover by ammonia recycle percolation process. Bioresour Technol 96(18):2007-2013

89. King JW, Srinivas K (2009) Multiple unit processing using sub- and supercritical fluids. J Supercrit Fluids 47(3):598-610

90. Kovacs K, Megyeri L, Szakacsa G, Kubicekc CP, Galbeb M, Zacchi G (2008) Trichoderma atroviride mutants with enhanced production of cellulase and β-glucosidase on pretreated willow. Enzyme Microb Technol 43:48-55

91. Kuhad RC, Gupta R, Khasa YP (2010) Bioethanol production from lignocellulosic biomass: an overview. In: Lal B (ed) Wealth from waste, Teri Press, New Delhi, India.

92. Kumar P, Barrett DM, Delwiche MJ, Stroeve P (2009) Methods for pretreatment of lignocellulosic biomass for efficient hydrolysis and biofuel production. Ind Eng Chem Res 48:3713-3729

93. Kumar R, Singh S, Singh OV (2008) Bioconversion of lignocellulosic biomass: biochemical and molecular perspectives. J Ind Microbiol Biotechnol 35:377-391

94. Kuo CH, Lee CK (2009) Enhanced enzymatic hydrolysis of sugarcane bagasse by N-methylmorpholine-N-oxide pretreatment. Bioresour Technol 100(2):866-871

95. Lachenmeier DW, Godelmann R, Steiner M, Ansay B, Weigel J, Krieg G (2010) Rapid and mobile determiantion of alcoholic stregth in wine, beer, and sprits using a flow-through infrared sensor. Chem Cent J 4:5-15

96.  Lavarack BP, Giffin GJ, Rodman D (2002) The acid hydrolysis of sugarcane bagasse hemicellulose to produce xylose, arabinose, glucose, and other products. Biomass Bioenerg 23:367-380

97.  Lelkes Z, Szitkai Z, Rev E, Fonyo Z (2000) Rigorous MINLP model for ethanol dehydration system. Computers and Chem Eng 24:1331-1336

98.  Li Q, He YC, Xian M (2009) Improving enzymatic hydrolysis of wheat straw using ionic liquid 1-ethyl-3-methyl imidazolium diethyl phosphate pretreatment. Bioresour Technol 100(14):3570-3575

99.  Li YL, Li DC, Teng FC (2006) Purification and characterization of a cellobiohydrolase from the thermophilic fungus Chaetomium thermophilus CT2. Wei Sheng Wu Xue Bao 46:143-146

100. Lin Y, Tanaka S (2006) Ethanol fermentation from biomass resources: current state and prospects. Appl Microbiol Biotechnol 69:627-642

101. Lu GQ, Huang HH, Zhang DP (2006) Application of near-infrared spectroscopy to predict sweet potato starch thermal properties and noodle quality. J. Zhejiang University (Eng. Sci.) 6:475-481

102. Lynd LR, van Zyl WH, McBride JE, Laser M (2005) Consolidated bioprocessing of cellulosic biomass: an update. Curr Opin Biotechnol 16:577-583

103. Lynd LR, Jin H, Michels JG, Wyman CE, Dale B (2003) Bioenergy: background, potential, and policy. Center for Strategic & International Studies, Washington, D.C

104. Lynd LR, Weimer PJ, van Zyl WH, Pretorius IS (2002) Microbial cellulose utilization: fundamentals and biotechnology. Microbiol Mol Biol Rev 66:506-577

105. MacDonald DG, Bakhshi N, Mathews JF, Roychowdhury A, Bajpai P, Moo-Young M (1983) Alkaline treatment of corn stover to improve sugar production by enzymatic hydrolysis. Biotechnol Bioeng 25(8):2067-2076

106. McIntosh S, Vancov T (2010) Enhanced enzyme saccharification of sorghum bicolor straw using dilute alkali pretreatment. Bioresour Technol 101(17):6718-6727

107. Maheshwari DK, Gohade S, Paul J, Verma A (1994) A paper mill sludge as a potential source for cellulase production by

Trichoderma reesei QM9123 and Aspergillus Niger using mixed cultivation. Carbohydr Polym 23:161-163

108. Maheshwari R, Bharadwaj G, Bhat MK (2000) Thermophilic fungi: their physiology and enzymes. Microbiol Mol Biol Rev 64:461-488

109. Mandels M, Weber J, Parizek R (1971) Enhanced cellulase production by a mutant of Trichoderma viride. Appl Microbiol 21:152-154

110. Martin C, Klinke HB, Thomsen AB (2007) Wet oxidation as a pretreatment method for enhancing the enzymatic convertibility of sugarcane bagasse. Enzyme and Microbial Technol 40(3):426-432

111. Martin C, Thomsen AB (2007) Wet oxidation pretreatment of lignocellulosic residues of sugarcane, rice, cassava and peanuts for ethanol production. J of Chemical Technol and Biotechnol 82(2):174-181

112. Martin C, Marcet M, Thomsen AB (2008) Comparison between wet oxidation and steam explosion as pretreatment methods for enzymatic hydrolysis of sugarcane bagasse. BioResources 3(3):670-683

113. Martinez AT, Speranza M, Ruiz-Duenas FJ, Ferreira P, Camarero S, Guillen F, Martinez MJ, Gutierrez A, del Rio JC (2005) Biodegradation of lignocellulosics: microbial, chemical, and enzymatic aspects of the fungal attack of lignin. Int Microbiol 8:195-204

114. McMillan JD (1994) Pretreatment of lignocellulosic biomass. In: Himmel ME, Baker JO, Overend RP (eds) Enzymatic conversion of biomass for fuels production, American Chemical Society, Washington, DC. pp 292-324

115. Meena K, Raja TK (2004) Immobilization of yeast invertase by gel entrapment. Ind J of Biotechnol 3:606-608

116. Millet MA, Baker AJ, Scatter LD (1976) Physical and chemical pretreatment for enhancing cellulose saccharification. Biotech Bioeng Symp 6:125-153

117. Montane D, Farriol X, Salvado J, Jollez P, Chornet E (1998) Application of stream explosion to the fractionation and rapid vapour-phase alkaline pulping of wheat straw. Biomass Bioenergy 14:261-276

118. Moredo N, Lorenzo M, Domínguez A, Moldes D, Cameselle C, Sanroman A (2003) Enhanced ligninolytic enzyme production and degrading capability of Phanerochaete chrysosporium and Trametes versicolor. World J Microb Biotechnol 19:665-669

119. Morohoshi N (1991) Chemical characterization of wood and its components. In: Hon DNS, Shiraishi N (eds) Wood and cellulosic chemistry, Marcel Dekker, Inc, New York, USA. pp 331-392

120. Mosier N, Wyman CE, Dale BD, Elander RT, Lee YY, Holtzapple M, Ladisch CM (2005) Features of promising technologies for pretreatment of lignocellulosic biomass. Bioresour Technol 96:673-686

121. Murray P, Aro N, Collins C, Grassick A, Penttila M, Saloheimo M, Tuohy M (2004) Expression in Trichoderma reesei and characterisation of a thermostable family 3 beta-glucosidase from the moderately thermophilic fungus Talaromyces emersonii. Protein Expr Purif 38:248-257

122. Ohta K, Beall DS, Meija JP, Shanmugam KT, Ingram LO (1991) Metabolic engineering of Klebsiella oxytoca M5A1 for ethanol production from xylose and glucose. Appl Environ Microbiol 57:2810-2815

123. Okada H, Mori K, Tada K, Nogawa M, Morikawa Y (2000) Identification of active site carboxylic residues in Trichoderma reesei endoglucanase Cel12A by site-directed mutagenesis. J Mol Catal 10:249-255

124. Palacios-Orueta A, Chuvieco E, Parra A, Carmona-Moreno C (2005) Biomass burning emissions: a review of models using remote-sensing data. Environ Monit Assess 104:189-209

125. Panda T, Bisaria VS, Ghose TK (1983) Studies on mixed fungal culture for cellulase and hemicellulase production. Part 1. Optimization of medium for the mixed culture of Trichoderma reesei DI-6 and Aspergillus wentii Pt 2804. Biotechnol Lett 5:767-772

126. Pappas C, Tarantilis PA, Daliani I, Mavromoustakos T, Polissiou M (2002) Comparison of classical and ultrasound-assisted isolation procedures of cellulose from kenaf (Hibiscus cannabinus L.) and eucalyptus (Eucalyptus rodustrus Sm.). Ultrason Sonochem 9(1):19-23

127. Pere J, Puolakka A, Nousiainen P, Buchert J (2001) Action of purified Trichoderma reesei cellulases on cotton fibers and yarn. J Biotechnol 89(2–3):247-255

128. Perepelkin KE (2007) Lyocell fibres based on direct dissolution of cellulose in N-methylmorpholine N-oxide: development and prospects. Fibre Chemistry 39(2):163-172

129. Perez VH, Reyes AF, Justo OR, Alvarez DC, Alegre RM (2007) Bioreactor coupled with electromagnetic field generator: effects of extremely Low frequency electromagnetic fields on ethanol production by Saccharomyces cerevisiae. Biotechnol Progress 23:1091-1094

130. Persson T, Matusiak M, Zacchi G, Jonsson A-S (2006) Extraction of hemicelluloses from process water from the production of masonite. Desalination 199:411-412

131. Philippidis GP, Smith TK, Wyman CE (1993) Study of the enzymatic hydrolysis of cellulose for production of fuel ethanol by the simultaneous saccharification and fermentation process. Biotechnol Bioeng 41:846-853

132. Qin Y, Wei X, Song X, Qu Y (2008) Engineering endoglucanase II from Trichoderma reesei to improve the catalytic efficiency at a higher pH optimum. J Biotechnol 135:190-195

133. Ragauskas AJ, Williams CK, Davison BH, Britovsek G, Cairney J, Eckert CA, Frederick WJ, Hallett JP, Leak DJ, Liotta CL, Mielenz JR, Murphy R, Templer R, Tschaplinski T (2006) The path forward for biofuels and biomaterials. Science 311:484-489

134. Rahman Z, Shida Y, Furukawa T, Suzuki Y, Okada H, Ogasawara W, Morikawa Y (2009) Application of Trichoderma reesei cellulase and xylanase promoters through homologous recombination for enhanced production of extracellular beta-glucosidase I. Biosci Biotechnol Biochem 73:1083-1089

135. Raposo S, Pardao JM, Diaz I, Costa MEL (2009) Kinetic modelling of bioethanol production using agro-industrial by-products. Int J of Energy Env 3(1):8

136. Rolz C (1986) Ultrasound effect of enzymatic saccharification. Biotech Letters 8(2):131-136

137. Rosenau T, Potthast A, Sixta H, Kosma P (2001) The chemistry of side reactions and byproduct formation in the system NMMO/cellulose (lyocell process). Prog Polym Sci 26(9):1763-1837

138. Rosgaard L, Pedersen S, Cherry JR, Harris P, Meyer AS (2006) Efficiency of new fungal cellulase systems in boosting enzymatic degradation of barley straw lignocellulose. Biotechnol Prog 22:493-498

139. Ruffell J, Levie B, Helle S, Duff S (2010) Pretreatment and enzymatic hydrolysis of recovered fibre for ethanol production. Bioresour Technol 101(7):2267-2272

140. Saha BC (2003) Hemicellulose bioconversion. J Ind Microbiol Biotechnol 30:279-291

141. Saha BC, Cotta MA (2007) Enzymatic saccharification and fermentation of alkaline peroxide pretreated rice hulls to ethanol. Enzyme Microbiol Technol 41:528-532

142. Saha BC, Cotta MA (2011) Continuous ethanol production from wheat straw hydrolysate by recombinant ethanologenic Escherichia coli strain FBR5. Appl Microbiol Biotechnol 90:477-487

143. Sakai Y, Tamiya Y, Takahashi F (1994) Enhancement of ethanol formation by immobilized yeast containing iron powder or Ba-ferrite due to eddy current or hysteresis. J Ferment Bioeng 77:169-172

144. Sang-Mok L, Koo YM (2001) Pilot-scale production of cellulase using Rut C-30 in fed-batch mode. J Microbiol Biotechnol 11(2):229-233

145. Sanchez C (2009) Lignocellulosic residues: biodegradation and bioconversion by fungi. Biotechnol Adv 27(2):185-194

146. Sasikumar E, Viruthagiri T (2010) Simultaneous saccharification and fermentation (SSF) of sugarcane bagasse - kinetics and modeling. Int J of Chemical and Biological Eng 3(2):57-64

147. Schmidt AS, Mallon S, Thomsen AB, Hvilsted S, Lawther JM (2002) Comparison of the chemical properties of wheat straw and beech fibers following alkaline wet oxidation and laccase treatments. J of Wood Chem and Technol 22(1):39-53

148. Schurz J (1978)Ghose TK (ed) Bioconversion of cellulosic substances into energy chemicals and microbial protein symposium proceedings, IIT, New Delhi. p 37

149. Scott WE, Gerber P (1995) Using ultrasound to deink xerographic waste. Tappi J 78:125-130

150. Senthilkumar V, Gunasekaran P (2005) Bioethanol production from cellulosic substrate: engineered bacteria and process integration challenges. J of Sci and Inds Resrch 64:845-853

151. Shallom D, Shoham Y (2003) Microbial hemicellulases. Curr Opin Microbiol 6:219-228

152. Sharma N, Kalra KL, Oberoi HS, Bansal S (2007) Optimization of fermentation parameters for production of ethanol from kinnow waste and banana peels by simultaneous saccharification and fermentation. Ind J Microbiol 47:310-316

153. Sharma K, Sharma SP, Lahiri SC (2009) Novel method for identification and quantitation of methanol and ethanol in alcoholic beverages by gas chromatography-Fourier transform infrared spectroscopy and horizontal attenuated total reflectance-Fourier transform infrared spectroscopy. J AOAC Int 92:518-526

154. Shafiei M, Karimi K, Taherzadeh MJ (2010) Pretreatment of spruce and oak by N-methylmorpholine-N-oxide (NMMO) for efficient conversion of their cellulose to ethanol. Bioresour Technol 101(13):4914-4918

155. Shafizadeh F, Bradbury AGW (1979) Thermal degradation of cellulose in air and nitrogen at low temperatures. J Appl Poly Sci 23:1431-1442

156. Shaw AJ, Podkaminer KK, Desai SG, Bardsley JS, Rogers SR, Thorne PG, Hogsett DA, Lynd LR (2008) Metabolic engineering of a thermophilic bacterium to produce ethanol at high yield. Proc Natl Acad Sci U S A 105(37):13769-13774

157. Shi J, Chinn MS, Sharma-Shivappa RR (2008) Microbial pretreatment of cotton stalks by solid state cultivation of phanerochaete chrysosporium. Bioresour Technol 99:6556-6564

158. Sills DL, Gossett JM (2011) Assessment of commercial hemicellulases for saccharification of alkaline pretreated perennial biomass. Bioresour Technol 102(2):1389-1398

159. Sims R (2003) Biomass and resources bioenergy options for a cleaner environment in developed and developing countries. Elsevier Science London, UK.

160. Sipos B, Reczey J, Somorai Z, Kadar Z, Dienes D, Reczey K (2009) Sweet sorghum as feedstock for ethanol production: enzymatic

hydrolysis of steam pretreated bagasse. Appl Biochem Biotechnol 153:151-162

161. Sipos B, Kreuger E, Svensson S-E, Reczey K, Bjornsson L, Zacchi G (2010) Steam pretreatment of dry and ensiled industrial hemp for ethanol production. Biomass Bioenergy.

162. Sohel MI, Jack MW (2010) Therrmodynamic analysis of lignocellulosic biofuel production via a biochemical process: guiding technology selection and research focus. Bioresour Technol.

163. Soloveva IV, Okunev ON, Velkov VV, Koshelev AV, Bubnova TV, Kondrateva EG, Skomarovskii AA, Sinitsyn AP (2005) The selection and properties of Penicillium verruculosum mutants with enhanced production of cellulases and xylanases. Mikrobiologiia 74:172-178

164. Soto ML, Dominguez H, Nunez MJ, Lema JM (1994) Enzymatic saccharification of alkali-treated sunflower hulls. Bioresour Technol 49(1):53-59

165. Stenberg K, Bollok M, Reczey K, Galbe M, Zacchi G (2000) Effect of substrate and cellulase concentration on simultaneous saccharification and fermentation of steam-pretreated softwood for ethanol production. Biotechnol Bioeng 68:204-210

166. Sternberg D (1976) Production of cellulase by trichoderma. Biotechnol Bioeng Symp 6:35-53

167. Stockton BC, Mitchell DJ, Grohmann K, Himmel ME (1991) Optimum β-D-glucosidase supplementation of cellulase for efficient conversion of cellulose to glucose. Biotechnol Lett 13:57-62

168. Sukumaran RK, Singhania RR, Pandey A (2005) Microbial cellulases—production, applications and challenges. Jr of Sci and Inds Resrch 64(11):832-844

169. Sun FB, Cheng HZ (2007) Evaluation of enzymatic hydrolysis of wheat straw pretreated by atmospheric glycerol autocatalysis. J Chem Technol Biotechnol 82:1039-1044

170. Sun RC, Tomkinson J (2002) Comparative study of lignins isolated by alkali and ultrasound-assisted alkali extractions from wheat straw. Ultrason Sonochem 9(2):85-93

171. Sun YE, Cheng J (2002) Hydrolysis of lignocellulosic materials for ethanol production: a review. Bioresour Technol 83(1):1-11

172. Swatloski RP, Spear SK, Holbrey JD, Rogers RD (2002) Dissolution of cellulose with ionic liquids. J Am Chem Soc 124:4974-4975

173. Szczodrak J (1989) The use of cellulases from a j-glucosidasc hyperproducing mutant of trichoderma reesei in simultaneous saccharification and fermentation of wheat straw. Biorechnol Bioeng 9:1112

174. Szijarto N, Kadar Z, Varga E, Thomsen AB, Costa-Ferreira M, Reczey K (2009) Pretreatment of reed by wet oxidation and subsequent utilization of the pretreated fibers for ethanol production. Appl Biochem Biotechnol 155(1–3):386-396

175. Szitkai Z, Lelkes Z, Rev E, Fonyo Z (2002) Optimization of hybrid ethanol dehydration systems. Chem Eng and Processing 41:631-646

176. Taherzadeh MJ, Karimi K (2008) Pretreatment of lignocellulosic wastes to improve ethanol and biogas production: a review. Int J Mol Sci 9:1621-1651

177. Takagi M, Abe S, Suzuki S, Emert GH, Yata N (1977) A method for production of alcohol directly from cellulose using cellulose and yeast. Proceedings, Bioconversion Symposium, IIT, Delhi. pp 551-571

178. Talebnia F, Karakashev D, Angelidaki I (2010) Production of bioethanol from wheat straw: an overview on pretreatment, hydrolysis and fermentation. Bioresour Technol 101(13):4744-4753

179. Teymouri F, Laureano-Perez L, Alizadeh H, Dale BE (2004) Ammonia fiber explosion treatment of corn stover. Appl Biochem and Biotechnol, Part A 115(1–3):951-963

180. Tsuyomoto M, Teramoto A, Meares P (1997) Dehydration of ethanol on a pilot plant scale, using a new type of hollow-fiber membrane. J of Membrane Sci 133:83-94

181. Turner P, Mamo G, Karlsson EN (2007) Potential and utilization of thermophiles and thermostable enzymes in biorefining. Microb Cell Fact 6:9

182. Viikari L, Alapuranen M, Puranen T, Vehmaanpera J, Siika-Aho M (2007) Thermostable enzymes in lignocellulose hydrolysis. Adv Biochem Eng Biotechnol 108:121-145

183. Vinatoru M, Toma M, Mason TJ (1999) Ultrasonically assisted extraction of bioactive principles from plants and their constituents. In: Mason TJ (ed) Advances in sonochemistry, 5th edn. JAI Press, London. pp 209-248

184. Wang ML, Choong YM, Su NW, Lee MH (2003) A rapid method for determination of ethanol in alcholic beverages using capillary Gas chromatography. Jr of Food and Drug Analysis 11(2):1330-140

185. Wang ZM, Li L, Xiao KJ, Wu JY (2009) Homogeneous sulfation of bagasse cellulose in an ionic liquid and anticoagulation activity. Bioresour Technol 100(4):1687-1690

186. Wati L, Kumari S, Kundu BS (2007) Paddy straw as a substrate for ethanol production. Ind Jr Of Microbiol 47:26-29

187. Wen Z, Liao W, Chen S (2004) Hydrolysis of animal manure lignocellulosics for reducing sugar production. Bioresour Technol 91:31-39

188. Wingren A, Galbe M, Zachhi G (2003) Techno-economic evaluation of producing ethanol from softwood: comparision of SSF and SHF and identification of bottle necks. Biotechnol Prog 19:1086-1093

189. Wood BE, Aldrich HC, Ingram LO (1997) Ultrasound stimulates ethanol production during the simultaneous saccharification and fermentation of mixed waste office paper. Biotechnol Prog 13(3):232-237

190. Wright JD (1998) Ethanol from biomass by enzymatic hydrolysis. Chem Eng Prog 84(8):62-74

191. Wyman CE, Spindler DD, Grohmann K, Lastick SM (1986) Simultaneous saccharification and fermentation of cellulose with the yeast Brettanomyces clausenii. Biotechnol and Bioeng Symp No 17:221-238

192. Wyman CE, Hinman ND (1990) Ethanol. Fundamentals of production from renewable feedstocks and use as a transportation fuel. Appl Biochem Biotechnol 24:735-753

193. Yamashita Y, Sasaki C, Nakamura Y (2010) Effective enzyme saccharification and ethanol production from Japanese cedar using various pretreatment methods. J Biosci Bioeng 110:79-86

194. Yang B, Wyman CE (2008) Pretreatment: The key to unlocking low cost cellulosic ethanol. Biofuel Bioprod Biorefin 2:26-40

195. Yu G, Yano S, Inoue H, Inoue S, Endo T, Sawayama S (2010) Pretreatment of rice straw by a hot-compressed water process for enzymatic hydrolysis. Appl Biochem Biotechnol 160(2):539-551

196. Zaldivar GJ, Nielsen J, Olsson L (2001) Fuel ethanol production from lignocellulose: a challenge for metabolic engineering and process integration. Appl Microbiol Biotechnol 56:17-34

197. Zhang L, Zhao H, Gan M, Jin Y, Gao X, Chen Q, Guan J, Wang Z (2011) Application of simultaneous saccharification and fermentation (SSF) from viscosity reducing of raw sweet potato for bioethanol production at laboratory, pilot and industrial scales. Bioresour Technol 102:4573-4579

198. Zhang M, Eddy C, Deanda K, Finkelstein M, Picataggio S (1995) Metabolic engineering of a pentose metabolism pathway in ethanologenic Zymomonas mobilis. Science 267:240-243

199. Zhang Y-HP, Lynd LR (2004) Toward an aggregated understanding of enzymatic hydrolysis of cellulose: non-complexed cellulase systems. Biotechnol Bioeng 88:797-824

200. Zhao H, Jones CL, Baker GA, Xia S, Olubajo O, Person VN (2009) Regenerating cellulose from ionic liquids for an accelerated enzymatic hydrolysis. J Biotechnol 139(1):47-54

201. Zhao Y, Wang Y, Zhu JY, Ragauskas A, Deng Y (2008) Enhanced enzymatic hydrolysis of spruce by alkaline pretreatment at low temperature. Biotechnol and Bioengi 99(6):1320-1328

202. Zhu JY, Pan XJ, Wang GS, Gleisner R (2009) Sulfite pretreatment for robust enzymatic saccharification of spruce and red pine. Bioresour Technol 100:2411-2418

203. Zhu S, Wu Y, Chen Q, Yu Z, Wang C, Jin S, Ding Y, Wu G (2006) Dissolution of cellulose with ionic liquids and its application: a mini-review. Green Chem 8:325-327

204. Zhu S, Wu Y, Yu Z, Zhang X, Wang C, Yu F, Jin S (2006) Production of ethanol from microwave-assisted alkali pretreated wheat straw. Process Biochem 41:869-873

# Chapter 6

# ScMT2-1-3, a Metallothionein Gene of Sugarcane, Plays an Important Role in the Regulation of Heavy Metal Tolerance/Accumulation

Jinlong Guo, Liping Xu, Yachun Su, Hengbo Wang, Shiwu Gao, Jingsheng Xu, and Youxiong Que

Key Lab of Sugarcane Biology and Genetic Breeding, Ministry of Agriculture, Fujian Agriculture and Forestry University, Fuzhou, Fujian 350002, China

## ABSTRACT

Plant metallothioneins (MTs), which are cysteine-rich, low-molecular-weight, and metal-binding proteins, play important roles in detoxification, metal ion homeostasis, and metal transport adjustment.

In this study, a novel metallothionein gene, designated as ScMT2-1-3 (GenBank Accession number JQ627644), was identified from sugarcane. ScMT2-1-3 was 700 bp long, including a 240 bp open reading frame (ORF) encoding 79 amino acid residues. A His-tagged ScMT2-1-3 protein was successfully expressed in Escherichia coli system which had increased the host cell's tolerance to $Cd^{2+}$, $Cu^{2+}$, PEG, and NaCl. The expression of ScMT2-1-3 was upregulated under $Cu^{2+}$ stress but downregulated under $Cd^{2+}$ stress. Real-time qPCR demonstrated that the expression levels of ScMT2-1-3 in bud and root were over 14 times higher than those in stem and leaf, respectively. Thus, both the E. coli assay and sugarcane plantlets assay suggested that ScMT2-1-3 is significantly involved in the copper detoxification and storage in the cell, but its functional mechanism in cadmium detoxification and storage in sugarcane cells needs more testification though its expressed protein could obviously increase the host E. coli cell's tolerance to $Cd^{2+}$. ScMT2-1-3 constitutes thus a new interesting candidate for elucidating the molecular mechanisms of MTs-implied plant heavy metal tolerance/accumulation and for developing sugarcane phytoremediator varieties.

# INTRODUCTION

Along with the population growth and the rapid development of industrialization and urbanization, our planet is constantly subjected to various kinds of pollution damage. The heavy metal-contaminated farmland in China had already topped 20 million hectares in 2003, accounting for about 1/5 of the total cultivated area [1]. Due to the heavy metal pollution, China's annual grain production cuts in more than 1,000 million tons, and this caused a direct economic loss of about 200 billion yuans [1]. The increasing trend of pollution continued from that time and is likely to continue over the next few decades if significant remedial measures are not implemented in China.

The concept of phytoremediation was first proposed by Chaney [2] and involved the use of plants to remove pollutants from the environment or to render them harmless [3]. Phytoremediation consists of mitigating pollutant concentrations in contaminated water, soils, or air, with the ability of plants to contain, degrade, or eliminate those materials of metals, pesticides, solvents, crude oil and its derivatives, explosives and various other contaminants from the media that contain them [4]. The plant-based remediation technologies have the potential to be low cost, low impact, visually benign, and environmentally sound [5]. In recent years, there has been an increasing interest in studying the molecular mechanisms of metal accumulation and tolerance in plants [6, 7].

Sugarcane (Saccharum spp. L.), a major sucrose accumulator and biomass producer, is one of the most important field crops grown in the tropics and subtropics. It accounts for more than 90% of China's total sugar output at present [8]. Due to its outstanding biomass production and economic importance, sugarcane offers the potential to be a phytoremediator species, while its prospective metal accumulation and tolerance have not been fully characterized. A research on this topic was carried out by Sereno et al. [9], which showed that sugarcane could be a copper (Cu) or cadmium (Cd) phytoremediator as its plantlets were able to tolerate up to $100\,\mu M$ of $Cu^{2+}$ or $500\,\mu M$ of $Cd^{2+}$ in nutrient solution for 33 days without symptoms of toxicity while accumulating $45\,mg\,Cu\,kg^{-1}$ or $451\,mg\,Cd\,kg^{-1}$ shoot dry weight, respectively, without significant reduction in fresh weight.

Metallothioneins (MTs) are cysteine-rich, low-molecular-weight, and metal-binding proteins, which have been found in a wide variety of organisms including animals, plants, cyanobacteria, and fungi [10]. Plant MTs are extremely diverse [11] and can be classified into four subfamilies (MT1 to MT4), based on the arrangement of Cys residues [10]. Due to their ability to reversibly bind both toxic and essential metal ions, plant MTs play important roles in detoxification, metal ion homeostasis, and metal transport adjustment [10]. Consequently, the role of plant MT genes in heavy metal tolerance mechanism and phytoremediation has attracted more and more attention in recent years [7], and their ability to metal accumulation and tolerance has been demonstrated in several plants. It was shown in Arabidopsis thaliana that expression of AtMT4a gene in vegetative tissues at different developmental stages conferred increased tolerance towards Cu and Zn [12]. Transgenic Avicennia marina that expresses AmMT2 have been scored for enhanced tolerance to Zn, Cd, Cu, and Pb [13].

Sugarcane is one of the few species that contain genes encoding all four types of MTs [10] and one of the most potential phytoremediation species for its outstanding biomass production and high metal enrichment capability. Undoubtedly, isolation and characterization of sugarcane MT genes from sugarcane are the basis for better understanding MT gene function in heavy metal tolerance mechanism and phytoremediation. In this study, an MT2 gene, termed ScMT2-1-3, was successfully isolated based on large sequencing and bioinformatics analysis of the sugarcane stem full-length cDNA library. ScMT2-1-3 protein was expressed in the E. coli Rosetta strain, and the transgenic bacteria showed an increased tolerance both to Cd and Cu. The expression patterns of ScMT2-1-3 in sugarcane plant were characterized in response to $CdCl_2$ and $CuCl_2$, and its expression levels in different sugarcane tissues were determined by real-time quantitative polymerase chain reaction (real-time qPCR).

# MATERIALS AND METHODS

Plant Materials and Treatment

Sugarcane varieties used in this study were provided kindly by the Key Laboratory of Sugarcane Biology and Genetic Breeding, Ministry of Agriculture (Fuzhou, China). Uniform tissue culture plantlets of an elite sugarcane variety FN39 were grown in 1/2 Hoagland nutrient solution for one week and then subjected to two different treatments: 500 µM $CdCl_2$ or 100 µM $CuCl_2$. The sampling times were 0 h, 3 h, 12 h, 48 h, and 72 h after the start of each treatment. All samples collected were immediately fixed in liquid nitrogen and stored in a refrigerator at −85°C until RNA extraction.

Nine healthy and consistent growing plants were randomly chosen and dug up with roots from sugarcane variety FN39 grown for 10 months. For each plant, the young root, maturing stem (internodes 4–6, 10–12, and 16–18), the leaf (+1), and all of the buds were sampled and fixed in liquid nitrogen. The collected materials were then stored in a refrigerator at −85°C until RNA extraction.

# Molecular Cloning, Sequencing, and Bioinformatics Analysis

The sugarcane stem full-length cDNA library was provided by the Key Laboratory of Sugarcane Biology and Genetic Breeding, Ministry of

Agriculture (Fuzhou, China). E. coli Rosetta (DE3) and the prokaryotic expression vector pET28a were purchased from Abmart, Inc. (Shanghai, China). The restriction enzymes EcoR I, Xho I, T4 DNA ligase, Ex-Taq enzyme, PrimeScript RT-PCR Kit, TaKaRa LA PCR in vitro Cloning Kit, DNA, and protein molecular weight (MW) markers were purchased from TaKaRa (Dalian, China). HisTrap HP column was purchased from GE Healthcare Life Sciences. RQ1 RNase-Free DNase was obtained from Promega Corporation (USA), SYBR Green PCR Master Mix Kit was purchased from Applied Biosystems (USA), and the instrument used in the real-time qPCR analysis was the ABI PRISM7500 real-time PCR system.

Large-scale sequencing and bioinformatics analysis of the full-length cDNA library of sugarcane stems were conducted as described by Guo et al. [14]. A full-length metallothionein homolog gene of sugarcane (namedScMT2-1-3) was identified by BLASTx (http://blast. ncbi.nlm.nih.gov/Blast.cgi) with a metallothio-2-superfamily domain (pfam01439). The open reading frame (ORF) of the full-length cDNA sequence ofScMT2-1-3 was predicted using the ORF finder online tool from NCBI (http://www.ncbi.nlm.nih.gov/gorf/gorf.html).

The accession numbers of the chosen proteins were AhMT2a (Arachis hypogaea) ABA08414, AhMT2b (A. hypogaea) ABB05520, AtMT2a (Arabidopsis thaliana) NP187550, AtMT2b (A. thaliana) NP195858 and AAA82212, NcMT2a (Noccaea caerulescens) ACR46970, NcMT2b (N. caerulescens) ACR46962, OsMT2a (Oryza sativa) P94029 and AAC49627, OsMT2b (O. sativa) A3AZ88 and AAB18814, OsMT2c (O. sativa) Q5JM82 and BAA19661, PMT2a (Populus trichocarpa × P. deltoids) AAT02524, PMT2b (P. trichocarpa × P. deltoids) AAt02525, PoMT2a (Posidonia oceanica) CAB96155, PoMT2b (P. oceanica) CAB96154, SbMT2 (Sorghum bicolor) XP002455197, SbMT2c (S. bicolor) XP002439147, SmMT2a (Salix matsudana) ABM21761, SmMT2b (S. matsudana) ABM21762, SnMT2a (Solanum nigrum) ACF10395, SnMT2b (S. nigrum) ACF10396, ScMT2-1-1 (Saccharum complex) (the deduced amino acid sequence of CA232620/ SCRUFL3063A10.g), ScMT2-1-2 (S. complex) AAV50043 and ABP37784, ScMT2-1-3 (S. complex) AFJ44225, ZmMT2-1 (Zea mays) NP001150795, and ZmMT2-2 (Z. mays) NP001147309. Alignment of putative ScMT2-1-3 protein sequence to MT2 proteins from A. hypogaea, A. thaliana, N. caerulescens, O. sativa, P. trichocarpa × P.

deltoids,P. oceanica, S. matsudana, S. nigrum, Z. mays, S. bicolor,and Saccharum complex was performed using DNAMAN 5.2.2 software.

# Plasmid Constructs

To study the function of ScMT2-1-3 in prokaryotes, the ScMT2-1-3 ORF with matched sites was amplified by PCR from the identified cDNA clone of the full-length cDNA library. The used PCR primer sequences were MT F: 5'-CGCGGATCCATG TCGTGCTGCGGAGGCAACTG-3' and MT R: 5'-CCGCTCGAGCTTGCAGGTG CAGGGGTTGCAGC-3' (BamH I and Xho I sites are underlined). PCR was performed in a reaction volume of 50 μL containing 5.0 μL 10× PCR buffer, 4.0 μL deoxynucleotide triphosphates (dNTPs) (2.5 mM), 2.0 μL each of forward and reverse primers (10 μM), 2.0 μL plasmid DNA (100 ng), 0.25 μL Ex-Taq enzyme (5 U/μL), and ddH$_2$O added as supplement. The PCR amplification program consisted of predenaturation for 5 min at 94°C, denaturation for 30 s at 94°C, annealing for 30 s at 55°C, extension for 30 s at 72°C for 30 cycles; and final extension for 10 min at 72°C. The ScMT2-1-3 ORF withBamH I and Xho I sites was subcloned into pET28a (+) (BamH I-Xho I sites) in the E. coli strain Rosetta to generate the putative recombinants. A bacterial clone containing the desired recombinant plasmid was identified and validated by PCR amplification, double digestion, and sequencing, and the clone was named as pET28a-MT2.

# SDS-PAGE and MALDI-TOF-TOF-MS Analysis of Prokaryotic Expression Products

The pET28a-MT2 and empty pET28a (+) were both transformed into E. coli Rosetta (DE3). The single colony was inoculated into an LB medium (20 mL) containing kanamycin (50 μg·mL$^{-1}$) and chloramphenicol (170 μg·mL$^{-1}$) and incubated with 150 rpm shaking overnight at 37°C. The following day, a dilution of 1% of this overnight cultured medium was inoculated into a fresh LB medium (20 mL) containing the same concentration of kanamycin and chloramphenicol and then shake-cultured in the same conditions. When OD$_{600}$ of the medium reached 0.4–0.6, a sample of 1.0 mL was collected as the control, and IPTG was then added to the remaining medium to a final concentration of

1.0 mM. The LB medium with pET28a-MT2(Rosetta) was induced for 2 h, 4 h, 6 h, and 8 h at 37°C. 100 μL of the medium was collected at each time point. LB media with empty pET28a (+) (Rosetta) and blank E. coli Rosetta were each induced in IPTG for 8 h, after which 100 μL of the cultures was collected and mixed with 25 μL 5× loading buffer and then heated at 100°C for 5 min. The 10 μL mixed sample was used for 12% SDS-PAGE loading. Protein molecular weight marker was used for monitoring protein separation during SDS-polyacrylamide gel electrophoresis. After electrophoresis, the gel was colored with coomassie brilliant blue and then imaged. Bio-Rad Quantity One 4.5.0 software was used to calculate the protein MW in SDS-polyacrylamide gel. At the same time, the theoretical MW of the recombinant protein was estimated using online software protein molecular weight (http://www.ualberta.ca/~stothard/javascript/protein_mw.html). The expression products were purified using HisTrap HP column, and the purified recombinant protein was analyzed by MALDI-TOF-TOF-MS for protein identification. The mass peak profiling was analyzed using online software Mascot (http://www.matrixscience.com/search_form_select.html) and MS-Digest (http://prospector.ucsf.edu/prospector/cgi-bin/msform.cgi?form=msdigest).

## Study on the Response of E. Coli Cells Containing Recombinant ScMT2-1-3 Gene to Abiotic Stresses

Spot assay was performed to ascertain the response of E. coli Rosetta (DE3) cells transformed with recombinant plasmid (pET28a-MT2) or vector alone (pET28a) to $Cd^{2+}$, $Cu^{2+}$, PEG, and NaCl stresses. When cells grew to 0.6 ($OD_{600}$) in LB medium, IPTG was added up to a final concentration of 1.0 mM, and then the cells were grown for further 12 h at 37°C. The cultures were diluted to 0.6 ($OD_{600}$) and then to $10^{-3}$ and to $10^{-4}$ [15]. In group one, 10 μL from each dilution was spotted on LB plates containing 100 μM, 250 μM, 500 μM, and 750 μM $CdCl_2$. In group two, 10 μL from each dilution was spotted on LB plates containing 50 μM, 100 μM, 250 μM, and 500 μM $CuCl_2$. In group three, 10 μL from each dilution was spotted on LB plates containing 250 mM, 500 mM, 750 mM, and 1000 mM NaCl. In group four, 10 μL from each dilution was spotted on LB plates infiltrated with 15.0%, 20.0%, 25.0%, 30.0%, and 35.0% PEG8000 [16, 17]. All the LB plates contained 50 μg·mL$^{-1}$ kanamycin and 170 μg·mL$^{-1}$ chloramphenicol.

## Expression Profile of ScMT2-1-3 under Heavy Metal Stresses

Total RNA isolation was performed using the TRIzol Reagent (Invitrogen). The removal of DNA from RNA samples was realized by RQ1 RNase-Free DNase (Promega). The reverse transcription was realized by following the specifications of the PrimeScript RT reagent Kit (TaKaRa). Finally, the real-time qPCR reaction was realized by using the SYBR Green PCR Master Mix (AB).

The 25S rRNA (BQ536525) and GAPDH (CA254672) genes were chosen as the internal control in the real-time qPCR analysis [18, 19], and the forward and reverse primers for 25S rRNA were 5'-GCAGCCAAGCGTTCATA GC-3' and 5'-CCTATTG GTGGGTGAACAATCC-3' and for GAPDH were 5'-CACGGCCACTGGAAGCA-3' and 5'-TCCTCAGGGTTCCTGATGCC-3' [18]. From the sequence of ScMT2-1-3, a pair of real-time qPCR primers was designed using the Primer Express 3.0 software, and the forward and reverse primers for ScMT2-1-3 were 5'- ACCACCCAGGCTCTCATC AT-3' and 5'- CACTTGCACCCGTCGTTC T-3', respectively.

The real-time qPCR reaction was realized with following conditions: 2 min at 50°C, 10 min at 95°C and then 40 cycles of 94°C for 15 s, and 60°C for 60 s. Each sample was repeated three times in the assay. When the reaction was completed, a melting curve was obtained. The $2^{-\Delta\Delta CT}$ method was adopted to analyze the real-time qPCR results [20].

# RESULTS AND ANALYSIS

## Cloning and Sequence Analysis of ScMT2-1-3

A full-length cDNA sequence of a metallothionein-like gene designated as ScMT2-1-3 (GenBank Accession number JQ627644) was obtained from sugarcane by large sequencing of a stem full-length cDNA library.

ScMT2-1-3 has a full length of 700 bp, with an ORF of 240 bp, 5' UTR (untranslated region) of 90 bp, and 3' UTR of 370 bp (Figure 1). The deduced protein of ScMT2-1-3 was a typical plant type 2 MT-like protein which contains 14 cysteine residues distributed in two conserved cysteine-rich domains. The N-terminal domain of ScMT2-1-3 formed by eight Cys, arranged as CC, CXC, CXC and CXXC motifs, and the C-terminal domain formed by three CXC motifs, where C stands for Cys and X for variable amino acids (Figure 1).

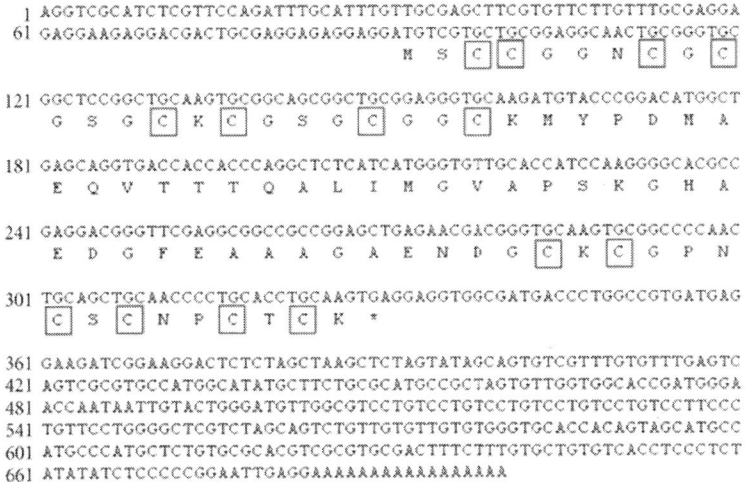

**Figure 1:** Nucleotide sequence and deduced amino acid sequence of ScMT2-1-3. Note: the C shows the conservative cysteine residual contained in two domains of ScMT2-1-3 with metal-binding motifs in combination among CC, CXC, and CXXC.

ScMT2-1-3 encodes a protein which is homologous to a metallothio-2-superfamily and contains two metal-binding domains (pfam01439). Twenty-four representative MT2 protein sequences from 11 plant species, including 8 MT2 defined originally as MT2a and 8 MT2 defined originally as MT2b from the same species, respectively, were chosen for analysis of their multiple alignments in this study. ScMT2-1-3 has high homology to other plant MTs and shared 95.00% and 93.83% identity with ScMT2-1-1 and ScMT2-1-2, respectively, by their protein sequences (Figure 2, Table 1).

**Table 1:** Classification of MT2s of plants

| Subgroup | Feature of Cys-rich domains at amino- and carboxy-terminal regions |
|----------|-------------------------------------------------------------------|
| MT2-1 | CCXXXCXCXXXCXCXXXCXXC......CXCXXXCXCXXCXC |
| MT2-2 | CCXXXCXCXXXCXCXXXCXXC......CXCXXCXCXXXCXCXCCXC |
| MT2-3 | CCXXXCXCXXXCXCXXXCXXC......CXCXXCXCXXXCXXCXCCXC |

Note: "C" represents a Cys residue, X represents an amino acid residue other than Cys, and "......" represents the intermediate region between the two Cys-rich domains.

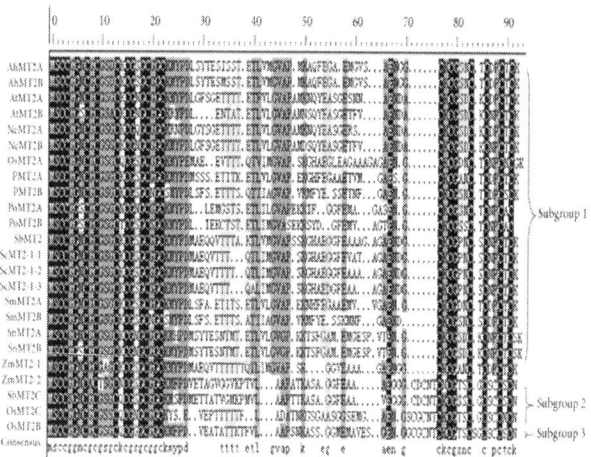

**Figure 2:** Amino acid sequence multiple alignments of MT2 proteins from different species.

# Expression of ScMT2-1-3 in E. Coli Rosetta

The recombinant protein was specifically induced after 2 h of IPTG treatment and reached a maximum at 8 h (Figure 3). The expression products were purified by HisTrap HP column and showed a single band when checking an SDS gel (Figure 4). The MW of recombinant protein (His-tagged-ScMT2-1-3) was estimated to be 12.34 kDa and 13.98 kDa using the online software MS-Digest and protein molecular weight, respectively, but the protein gave a 19.01 kDa band in the SDS gel when calculated by Quantity One 4.5.0 software (Bio-Rad). We repeated the experiment and got the same results, even by changing the E. coli host cell for BL21 (data not shown).

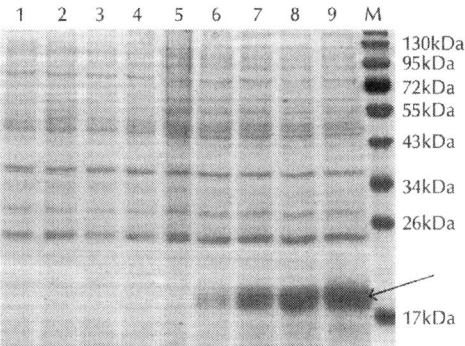

**Figure 3:** Protein expression of pET28a-MT2 in E. coli Rosetta strain. M, protein marker; 1, blank without induction; 2, blank induction for 8 h; 3, control without induction; 4, control induction for 8 h; 5, pET28a-MT2 without induction; 6 to 9, pET28a-MT2 induction for 2 h, 4 h, 6 h, and 8 h, respectively. IPTG-induced proteins shown by arrow.

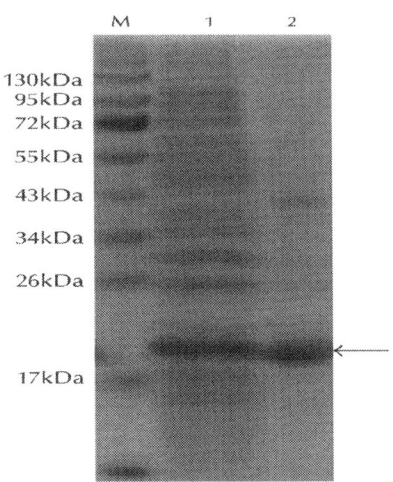

**Figure 4:** Protein purification of the recombinant protein. M, protein marker; 1, unpurified total protein; 2, Purified His-tagged-ScMT2-1-1 protein (shown by arrow).

Further validation of the recombinant protein was realized by using MALDI-TOF-TOF MS method, and the results were analyzed using online software Mascot and MS-Digest. Three mass peaks with value of 1 083.514, 1 535.647 and 1 768.869 (Figure 5) were matched with the peptide sequences of "LEHHHHHH," "GSHMASMTGGQQM GR," and "GSSHHHHHHSSGLVPR," respectively, which were partial sequences of the recombinant protein (Figure 6

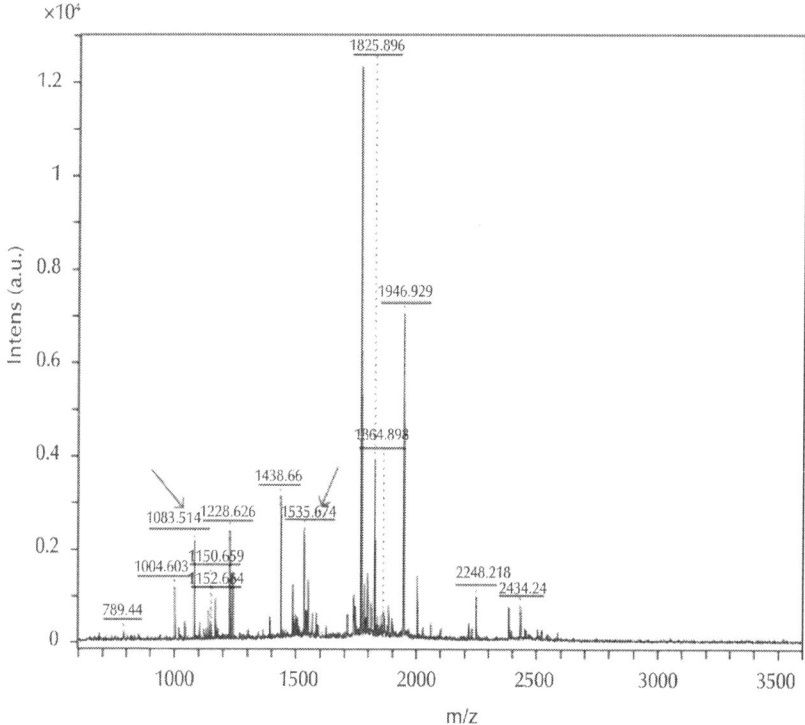

(a)

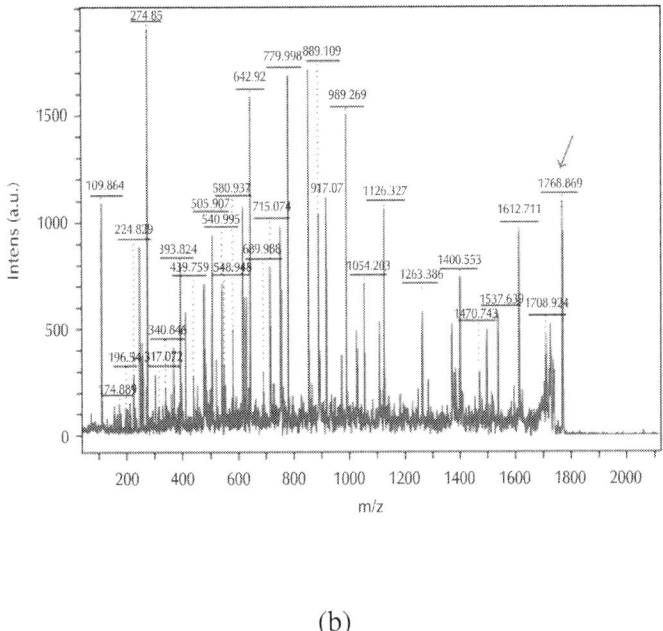

(b)

**Figure 5:** MALDI-TOF-TOF MS results of ScMT2-His.

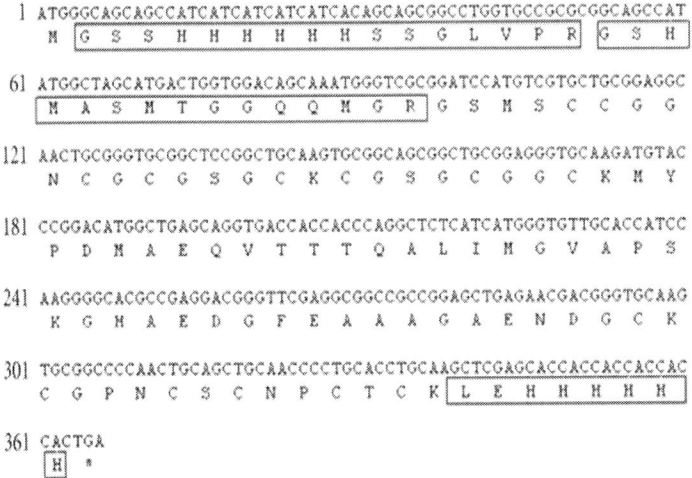

**Figure 6:** Nucleotide sequences of ScMT2-His and deduced amino acid sequences.

# Overexpression of ScMT2-1-3 in E. Coli Enhances Its Growth under Abiotic Stresses

Both the ScMT2-1-3 transformed and the control cells could grow in the plates containing $Cd^{2+}$, $Cu^{2+}$, and PEG, respectively. However, the former formed more colonies compared with the latter (Figures 7, 8, and 9). The results show that the recombinant protein enhances its growth under $Cd^{2+}$, $Cu^{2+}$, and PEG stresses. The growth difference was observed with the NaCl-containing LB plates after overnight culture. The ScMT2-1-3 expressed cells were able to tolerate high salt concentrations of up to 500 mM NaCl. In contrast, the growth of the control cells was obviously inhibited at 250 mM NaCl and completely inhibited at 500 mM NaCl, a lethal level for the control cells (Figure 10).

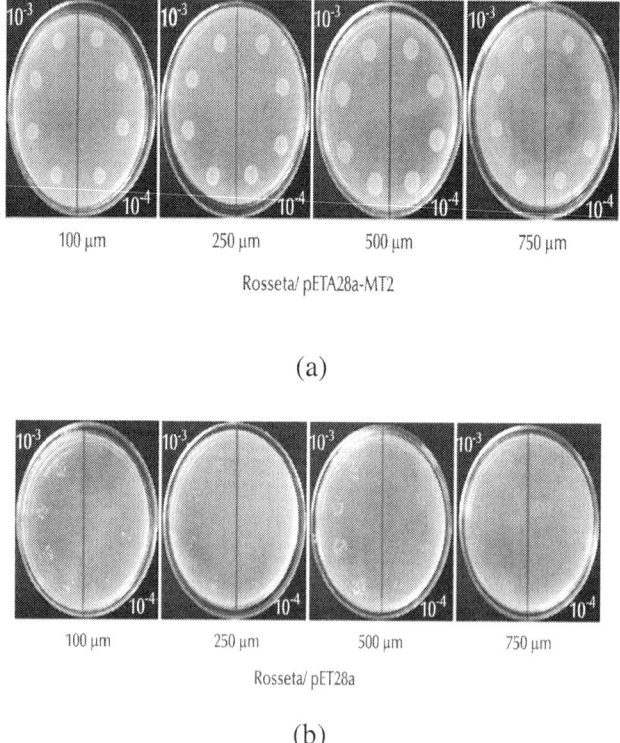

Rosseta/ pETA28a-MT2

(a)

Rosseta/ pET28a

(b)

**Figure 7:** Spot assay of Rosetta/pET28a and Rosetta/pET28a-MT2 on LB plates with $CdCl_2$.

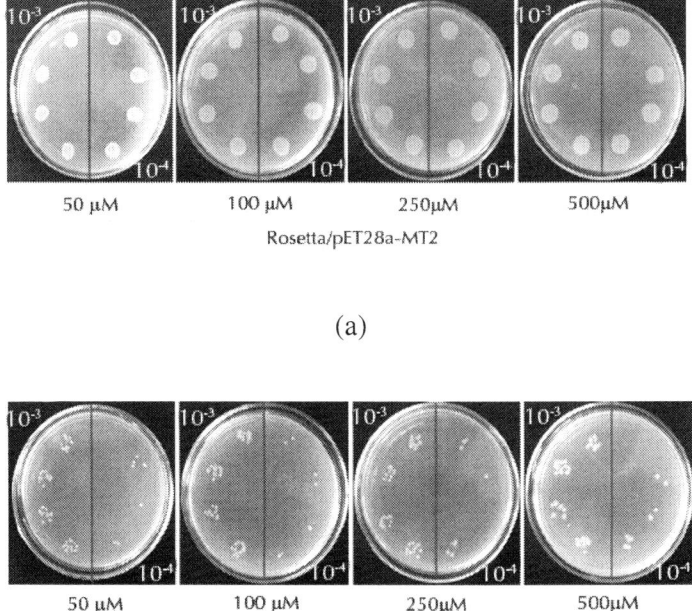

(a)

(b)

**Figure 8:** Spot assay of Rosetta/pET28a and Rosetta/pET28a-MT2 on LB plates with $CuCl_2$.

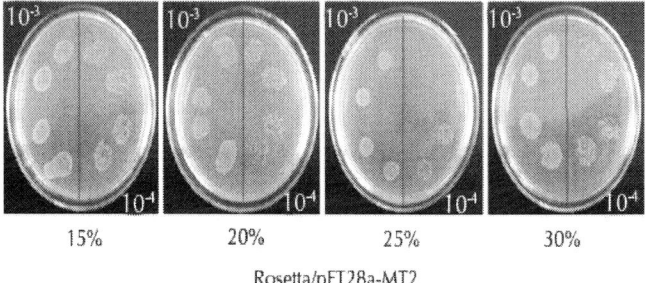

(a)

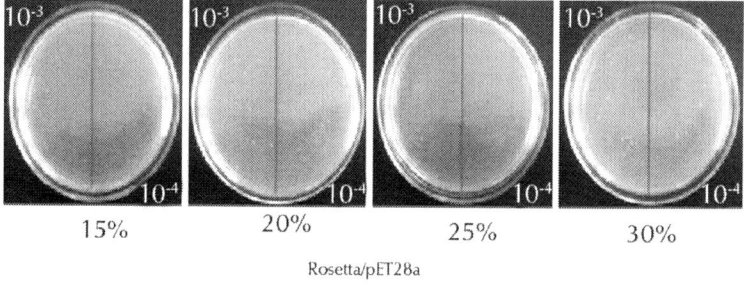

15%      20%      25%      30%

Rosetta/pET28a

(b)

**Figure 9:** Spot assay of Rosetta/pET28a and Rosetta/pET28a-MT2 on LB plates soaking with PEG.

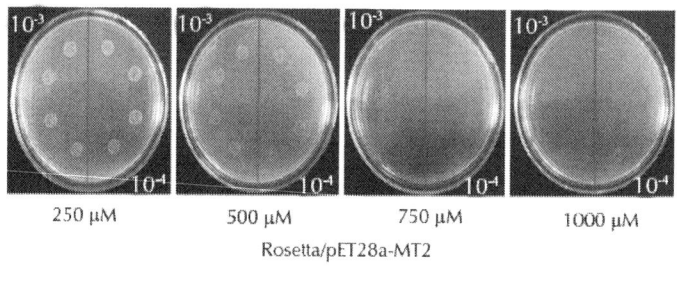

250 μM      500 μM      750 μM      1000 μM

Rosetta/pET28a-MT2

(a)

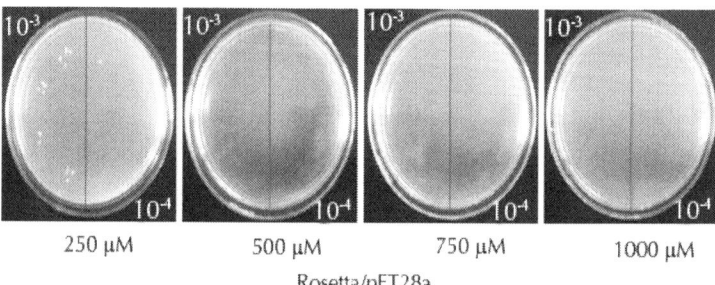

250 μM      500 μM      750 μM      1000 μM

Rosetta/pET28a

(b)

**Figure 10:** Spot assay of Rosetta/pET28a and Rosetta/pET28a-MT2 on LB plates with NaCl.

## Tissue-Specific Expression Analysis of ScMT2-1-3

For tissue-specific expression analysis of ScMT2-1-3, the sugarcane variety FN39 was used as experimental material, and the GAPDH gene was used as an internal control for real-time qPCR. The results showed that theScMT2-1-3 is highly expressed in root and bud but very lowly expressed in stem and leaf (Figure 11).

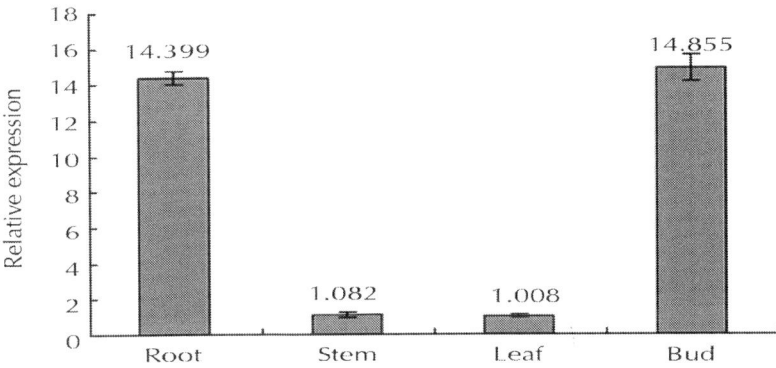

**Figure 11:** Tissue-specific expression analysis of the ScMT2-1-3 in sugarcane. Each value is the average of three replicate experiments ± standard error (n=3).

## Expression Profile of ScMT2-1-3 under Different Heavy Metal Stresses

Real-time qPCR was used to examine the expression profile of ScMT2-1-3 on sugarcane plantlets of the variety FN39 under $Cd^{2+}$ and $Cu^{2+}$ stresses, respectively. The real-time qPCR results showed that the expression ofScMT2-1-3 was inhibited by $Cd^{2+}$ stress which was visibly downregulated at 3 h following the treatment and maintained at a relatively lower level, compared to that of the control (Figure 12). In contrast, the expression ofScMT2-1-3 was upregulated after $Cu^{2+}$ treatment: first slightly increased at 3 h then obviously upregulated and reached its highest level at 12 h (2.87 times higher than that of the control), and the expression was maintained to a relatively high level (more than two times higher than that of the control) during the following examined time points (Figure 13).

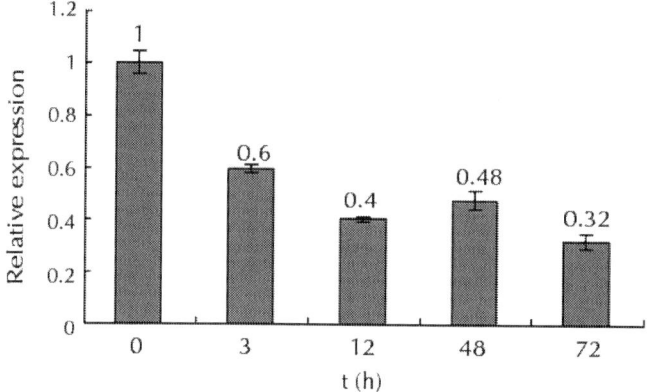

**Figure 12:** The ScMT2-1-3 expression in sugarcane under $CdCl_2$ stress. Each value is the average of three replicate experiments ± standard error (n=3).

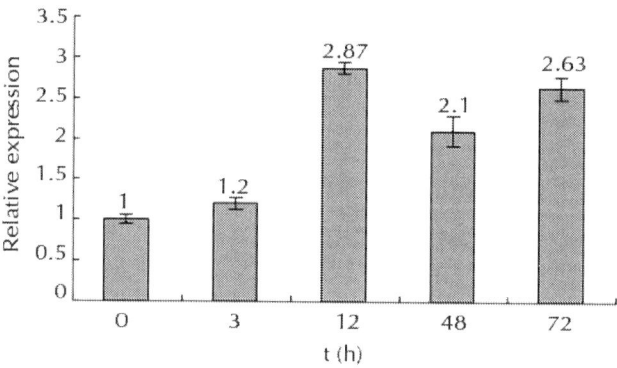

**Figure 13:** The ScMT2-1-3 expression in sugarcane under $CuCl_2$ stress. Each value is the average of three replicate experiments ± standard error (n=3).

# DISCUSSION

Plant metallothionein was first discovered from soybean in 1977 [21]. Based on the sequence homology, this family of genes can be grouped into four subfamilies (Type 1 to Type 4 or MT1 to MT4) [10, 11]. Considering the large member of the plant MT family and the high sequence diversity, further subdivision should be necessary for plant MTs. In A. thaliana, Zhou and Goldsbrough [22] classified the MT2 proteins into two subgroups, MT2a and MT2b, according to the four codons in the central domain of AtMT2a (codons 30–33: GFSG) which are absent in AtMT2b (Figure 2). This region was shown to be highly variable among plant MTs [22]. Our analysis in Figure 2 suggests that plant MT2 proteins can be subdivided into at least 3 subgroups according to the arrangement of Cys residues (Figure 2; Table 1). This is in agreement with Wong et al. [23] who classified the rice MT2s into three subgroups termed as OsMT2a, OsMT2b, and OsMT2c, respectively. In this study, the three subgroups were termed as MT2-1, MT2-2, and MT2-3, respectively.

From the alignment of sequences, we find that the sequences of the N-terminal domain of MT2 are highly conserved (MSCCGGNCGCS) (Figure 2). MT2-1 seems to be the most abundant class among the MT2 family with typical plant MT2 Cys-rich domains pattern characterized by Cobbett and Goldsbrough [10] (Figure 2). MT2-1 contains two cysteine-rich domains separated by a spacer of approximately 40 amino acid residues (Figure 2; Table 1). The N-terminal domain contains four Cys-containing motifs. The first pair of cysteines is present as a Cys-Cys motif in amino acid positions 3 and 4 of these proteins. Two Cys-Xaa-Cys motifs are present in the center of the N-terminal cysteine-rich domain. A Cys-Xaa-Xaa-Cys motif is at the end of the N-terminal cysteine-rich domain. The C-terminal domain contains three Cys-Xaa-Cys motifs. In this study, in addition to the highly conserved N-terminal domain of MT2 (MSCCGGNCGCS-) [10], we identified an additional highly conserved motif of GVAP among this subgroup (Figure 2).

Subgroups MT2-2 and MT2-3 have the same Cys-containing motif pattern in the N-terminal, but one additional Cys-Xaa-Cys motif is present at the beginning of the C-terminal cysteine-rich domain, and one more Cys residue is randomly present between the first two Cys-Xaa-Cys motifs and the last two Cys-Xaa-Cys motifs. Moreover, the last two Cys-Xaa-Cys motifs are arranged in tandem at the end of C-terminal (Figure 2; Table 1).

Sugarcane is one of the few species which contain genes encoding all four types of MTs [10]. Of the 291, 689 ESTs in the sugarcane expressed sequence tag (SUCEST) database; a total of 849 reads (0.29%) were found to encode metallothionein-like proteins and give 55 clusters which were conceptually translated and contained the full-length protein [24]. Among the 55 clusters, 21 were related to MT2 proteins and represented 8 protein sequence variants with minor amino acid changes [24]. When ScMT2-1-1 (SCRUFL3063A10.g/ CA232620) [9,24], ScMT2-1-2 (AAV50043 and ABP37784), ScMT2-1-3 (AFJ44225), and other 7 ScMT2 proteins [24] were compared by sequence alignment, we conclude that all the MT2s in sugarcane belong to the MT2-1 subgroup and share over 93% identity in their amino acid sequences (data not shown). To date, MT2-2 and MT2-3 subgroup types have not yet been reported in sugarcane.

The expression profile of MT2-1 genes in different organs, such as root, stem, and leaf, has been studied in several plant species. As a general evidence, the expression level of MT2-1 genes tends to be higher in leaves than that in roots [10, 22, 25, 26]. Both AtMT2a (CAA44630) and AtMT2b (AAA82212) from A. thaliana were found to be highly expressed in leaves but lowly expressed in roots from mature plants [22, 25]. A similar result was obtained for OsMT-2 gene (AAC49627) in rice [27]. In Avicennia marina, the level of expression of the gene AmMT2 in leaves was found to be over 1 times higher than that in stems and 2.1 times higher than that in roots [13]. In Hevea brasiliensis,the gene HbMT2 was also strongly expressed in leaves and in latex, but weakly in roots and in barks [28]. AmMT2 and HbMT2 were both classified into the subgroup of MT2-1 and shared 61.73% and 65.00% identities with ScMT2-1-3, respectively (data not shown). Based on the large-scale EST sequencing databases, the expression patterns of four types of MTs in sugarcane were investigated using 13 different sources of cDNA libraries including shoot-root zone, root, lateral bud, stem bark, stem internode, leaf, leaf roll, apex, flower, seed, callus, in vitro plantlet infected with Herbaspirillum rubri ssp. Albicans, and in vitro plantlets infected with Gluconsugarcane diazotroficans [24]. In general, the expression of MT2-1 in sugarcane tends to be lower in roots, higher in leaves, and so forth [9, 10, 22, 25, 26]. It is interesting to note that the expression level of ScMT2-1-3 in roots and in buds was significantly higher (over 14 times) than those in stems and in leaves (Figure 11), never reported before.

Difficulties in identifying and isolating MTs in plants may arise from the instability of these proteins in the presence of oxygen [10]. There were few reports about expressing plant MTs in prokaryotic system, though the research on plant MTs has been carried out for decades. In some earlier studies, plant MTs have been expressed in E. coli as GST fusions to examine the metal-binding properties of these proteins [13, 29, 30]. Recombinant production of MTs helps to circumvent some of the problems associated with direct isolation, and expression as a GST fusion offers simple possibilities for purification, quantification, and detection [30]. GST is commonly used to create fusion proteins, and many commercially available sources of GST-tagged plasmids include a thrombin domain for cleavage of the GST tag during protein purification. GST tag has the size of 220 amino acids (roughly 26 kDa), which, compared to the low molecular mass target protein MT, is quite big. Thus, the small His-tag may be a better choice when the function of fusion protein was studied in vivo. In the present study, His-tag fusion protein of His-ScMT2-1-3 had successful expression in E. coli Rosetta (DE3), and the His-ScMT2-1-3 has an observed MW which was much greater (5.03 kDa–6.67 kDa) than that predicted by their sequences. It has been reported that the basic amino acid residues of His-tag may retard the mobility of the fusion protein bands in SDS-PAGE and cause deviation in MW determination [31]. This deviation was not observed on GST-tag fusion proteins [13, 29, 30]. Though the MW of GST-tag fusion protein GST-AtMT2a was consistent with its predicted value, the value of thrombin cleavage product after removal of GST by affinity purification was estimated at least 15 kDa in SDS-PAG which was 3 kDa greater than the predicted one [29]. Thus, similarly, the electrophoretic mobility deviation was also observed in AtMT2a [29], and the deviation can be offset by the GST-tag for its 26 kDa MW which was much greater than AtMT2a (12 kDa). We infer that this deviation might be related to the characteristics of cysteine-rich.

AtMT2a (X62818) gene from A. thaliana has been shown to be able to complement the MT-deficiency in yeast (cup1), conferring a high level of resistance to $CuSO_4$ and a moderate resistance to $CdSO_4$ [22]. Guo et al. [26] have demonstrated that all four types of A. thaliana MTs, including AtMT2a (X62818) and AtMT2b(u11256), can offer a metal tolerance when expressed in Saccharomyces cerevisiae. Expression of the MT2-1 gene PsMTa (Z23097) from Pisum sativum in E. coli led to increased tolerance to copper and cadmium [32,33]. Overexpression of AmMT2 in E. coli BL (DE3) led to increased metal tolerance towards Zn, Cu, Pb, and Cd [13]. In a similar way, the expression of ScMT2-1-3 in E. coli Rosetta (DE3) enhances significantly the Cd and Cu tolerance in the present study. Furthermore, it leads to an increased tolerance to abiotic stresses of drought and salt.

Plant MTs exhibit beneficial metal-binding and induction properties which should protect these organisms from elevated levels of toxic heavy metals (such as Cd or Hg) and also affect, for example, the homeostasis of Cu and Zn, essential micronutrients for a range of plant physiological processes [10]. Some of the plant MTs' biological function of metal tolerance has been demonstrated in nonplant systems; however, MTs' in vivofunction in plants has not yet been elucidated. We take the MT2-1 homologous genes from various plants as samples in the following discussion. Using northern blotting technique, Zhou and Goldsbrough [22] had demonstrated that AtMT2a mRNA was present at a low level in A. thaliana 7-day-old seedlings, but the level of AtMT2a mRNA was increased in seedlings treated with $CuSO_4$ or $CdSO_4$ for 30 h. Moreover, this increase was positively correlated with metal concentration and exposure time [22]. Similarly, the regulation ofAmMT2 expression by Zn, Cu, or Pb was strongly dependent on the concentration and the time of exposure, as measured by real-time qPCR in seedlings of A. marina [13]. Conversely, the level of OsMT2a mRNA (u43530) from rice suspension cells was slightly reduced in the presence of excess Cd or Cu in the culture medium [27]. Exposures of 72 h to various concentrations of Cu, Cd, or Zn did not significantly affect the expression levels of TcMT2 in shoots of 5-week-old Thlaspi caerulescens seedlings [11]. A subsequent study ofAtMT2 on 7-day-old A. thaliana seedlings had demonstrated that AtMT2a is strongly induced by $CuSO_4$ (50 μM), whereas AtMT2b remains insensitive to the same condition [22]. It seems that TcMT2 and AtMT2b genes are expressed constitutively in some plant organs or tissues [11, 22]. Further study by real-time qPCR showed that although copper treatment (40 μM $CuCl_2$) failed to cause a significant increase in the expression ofAtMT2a in roots and in primary leaves of 6.5-day-old seedlings, the copper-induced increase in AtMT2amRNA was restricted to the cotyledons and, to a lesser extent, the hypocotyl [34]. Consistent with the results of García-Hernández et al. [34], RNA blots showed that the levels of AtMT2a and AtMT2b RNA increased after Cu treatment, but not for every gene in every tissue [25]. The Cu treatment increased the mRNA expression of AtMT2b in roots and AtMT2a in leaves [25]. Thus, they suggested that the plant MTs have distinct functions in heavy metal homeostasis [25]. It should be stressed that although it is believed that plant MTs could play an important role in heavy metal tolerance mechanism and phytoremediation, the precise function of these MTs in plant tolerance to abiotic stresses is still not clear because of the lack of information.

It has been reported that sugarcane plantlets were able to tolerate up to 100 mM of Cu or 500 mM of Cd in nutrient solution for 33 days while accumulating 45 mg Cu per kg or 451 mg Cd per kg shoot dry weight [9]. Using RNA blot, the expression patterns of sugarcane MT genes, including ScMT2-1-1, in shoots and in roots, were analyzed under increasing concentrations of copper and cadmium [9]. Increasing Cu concentration had little or no effect in modulating the expressions of MT genes, while an apparent minor modulation of some of the MT genes was detected in Cd treatments which presented a minor downregulation in 33 days Cd treatment samples. In this study, we showed that the level of ScMT2-1-3 expression in Cd-treated plantlets decreased steadily 3 h following the treatment and maintained a low expression level up to 72 h. This result was in agreement with Sereno et al. [9], who inferred that cadmium tolerance and accumulation in sugarcane might derive from other mechanisms. We infer that not ScMT2-1-3 but other member(s) of metallothioneins or phytochelatins play a key role in cadmium detoxification and homeostasis in sugarcane, although ScMT2-1-3 has the ability of imparting Cd tolerance when expressed in E. coli. Clearly different from ScMT2-1-1 observed by Sereno et al. [9], steadily the increased expression level of ScMT2-1-3 began to be observed at 3 h after Cu treatment, and the expression maintained 2 times higher than the control during the time examined. Thus, both the E. coli assay and sugarcane plantlets assays suggested that ScMT2-1-3 is significantly involved in the copper detoxification and storage in the cell. The differential expression patterns of ScMT2-1 in response to Cd or Cu exposure, observed by Sereno et al. [9], and this study, suggested that the members of ScMT2-1 genes may have diverse roles or functions.

According to their chemical and physical properties, two different molecular mechanisms of heavy metal toxicity caused by copper and cadmium have been reported: (a) production of reactive oxygen species by autoxidation and Fenton reaction, which is typical for transition metal copper [35, 36]; (b) blocking of essential functional groups in biomolecules, which is well documented for nonredox-reactive heavy metal cadmium [37]. On the one hand, the different expression pattern of ScMT2-1-3 may suggest different molecular mechanisms of heavy metal toxicity caused by $Cd^{2+}$ and $Cu^{2+}$ according to their chemical and physical properties. On the other hand, the up-regulation of ScMT2-1-3 under the stress of $Cu^{2+}$ indicated that this gene is significantly involved in the copper detoxification and storage in sugarcane cells, while the downregulation of ScMT2-1-3 under the stress of $Cd^{2+}$ implied that its functional mechanism in cadmium detoxification and storage in sugarcane cells needs more testification.

# CONCLUSIONS

In conclusion, we reported here a new member of plant type 2 metallothionein subfamily, termed as ScMT2-1-3 identified in sugarcane. We demonstrated that the expression of ScMT2-1-3 in E. coli can significantly enhance the tolerance to abiotic stresses such as heavy metal (copper and cadmium), droughtly and salt stresses. In contrast with the previous, reported MTs in sugarcane, ScMT2-1-3 has a distinct expression pattern in response to copper and cadmium treatments: highly expressed in root and bud but lowly expressed in stem and leaf; more interestingly, its expression is clearly upregulated by copper and downregulated by cadmium in sugarcane. These results, taken together, showed that ScMT2-1-3 was involved in the response to copper stresses, while cadmium tolerance and accumulation in sugarcane might derive from other mechanisms, maybe compensation mechanisms though this deduction needs more testification. ScMT2-1-3 constitutes thus a new interesting candidate for elucidating the molecular mechanisms of MTs-implied plant heavy metal tolerance/accumulation and for developing sugarcane phytoremediator varieties.

# ACKNOWLEDGMENTS

This work was supported by the earmarked fund for the Modern Agroindustry Technology Research System (CARS-20), the Natural

Science Foundation of Fujian province, China (Grant no. 2012J01089), Young Teacher Scientific Research Fund of Fujian Agriculture and Forestry University, China (Grant no. 2009002), Research Funds for Distinguished Young Scientists in Fujian Agriculture and Forestry University (xjq201202), and National High Technology Research and Development Program of China (863 Program) Project (2013AA102604).

# REFERENCES

1.  J. G. Gu, Q. X. Zhou, and X. Wang, "Reused path of heavy metal pollution in soils and its research advance," Journal of Basic Science and Engineering, vol. 11, no. 2, pp. 143–151, 2003.

2.  R. L. Chaney, "Plant uptake of inorganic waste constituents-," in Land Treatment of Hazardous Wastes, J. F. Parr, P. B. Marsh, and J. M. Kla, Eds., pp. 50–76, Noyes Data Corp., Park Ridge, NJ, USA.

3.  I. Raskin, R. D. Smith, and D. E. Salt, "Phytoremediation of metals: using plants to remove pollutants from the environment," Current Opinion in Biotechnology, vol. 8, no. 2, pp. 221–226, 1997.

4.  P. E. Flathman and G. R. Lanza, "Phytoremediation: current views on an emerging green technology," Journal of Soil Contamination, vol. 7, no. 4, pp. 415–432, 1998.

5.  S. D. Cunningham and D. W. Ow, "Promises and prospects of phytoremediation," Plant Physiology, vol. 110, no. 3, pp. 715–719, 1996.

6.  N. Verbruggen, C. Hermans, and H. Schat, "Molecular mechanisms of metal hyperaccumulation in plants," New Phytologist, vol. 181, no. 4, pp. 759–776, 2009.

7.  M. A. Hossain, P. Piyatida, J. A. Teixeira da Silva, and M. Fujita, "Molecular mechanism of heavy metal toxicity and tolerance in plants: central role of glutathione in detoxification of reactive oxygen species and methylglyoxal and in heavy metal chelation," Journal of Botany, vol. 2012, Article ID 872875, 37 pages, 2012.

8.  M. G. Wu, Y. Q. Lin, and H. Zhang, "Research status and prospect on industrial standard of sugarcane in China," Subtropical Agriculture Research, vol. 6, no. 3, pp. 209–211, 2010.

9.  M. L. Sereno, R. S. Almeida, D. S. Nishimura, and A. Figueira, "Response of sugarcane to increasing concentrations of copper

and cadmium and expression of metallothionein genes," Journal of Plant Physiology, vol. 164, no. 11, pp. 1499–1515, 2007.

10. C. Cobbett and P. Goldsbrough, "Phytochelatins and metallothioneins: roles in heavy metal detoxification and homeostasis," Annual Review of Plant Biology, vol. 53, pp. 159–182, 2002.

11. N. H. Roosens, R. Leplae, C. Bernard, and N. Verbruggen, "Variations in plant metallothioneins: the heavy metal hyperaccumulator Thlaspi caerulescens as a study case," Planta, vol. 222, no. 4, pp. 716–729, 2005.

12. I. D. Rodríguez-Llorente, "Epxression of the seed-specific metallothionein mt4a in plant vegetative tissues increases Cu and Zn tolerance," Plant Science, vol. 178, no. 3, pp. 327–332, 2010.

13. G. Y. Huang and Y. S. Wang, "Expression and characterization analysis of type 2 metallothionein from grey mangrove species (Avicennia marina) in response to metal stress," Aquatic Toxicology, vol. 99, no. 1, pp. 86–92, 2010.

14. J. L. Guo, Y. X. Que, J. X. Liu, Y. F. Zheng, R. K. Chen, and L. P. Xu, "Construction of full-length cDNA library for sugarcane stem by optimized oligo-capping," Chinese Journal of Troical Crops, vol. 30, no. 5, pp. 672–676, 2009.

15. K. Gupta, P. K. Agarwal, M. K. Reddy, and B. Jha, "SbDREB2A, an A-2 type DREB transcription factor from extreme halophyte Salicornia brachiata confers abiotic stress tolerance in Escherichia coli," Plant Cell Reports, vol. 29, no. 10, pp. 1131–1137, 2010.

16. C. M. van der Weele, W. G. Spollen, R. E. Sharp, and T. I. Baskin, "Growth of Arabidopsis thalianaseedlings under water deficit studied by control of water potential in nutrient-agar media," Journal of Experimental Botany, vol. 51, no. 350, pp. 1555–1562, 2000.

17. L. J. Zhang, L. J. Huan, Y. Y. Ruan, and Y. X. Guan, "Application of polyethylene glycol in the study of plant osmotic stress physiology," Plant Physiology Communications, vol. 40, pp. 361–364, 2004.

18. H. M. Iskandar, R. S. Simpson, R. E. Casu, G. D. Bonnett, D. J. Maclean, and J. M. Manners, "Comparison of reference genes for quantitative real-time polymerase chain reaction analysis of gene

expression in sugarcane," Plant Molecular Biology Reporter, vol. 22, no. 4, pp. 325–337, 2004.

19.  Y. X. Que, L. P. Xu, J. S. Xu, J. S. Zhang, M. Q. Zhang, and R. K. Chen, "Selection of control genes in real-time qPCR analysis of gene expression in sugarcane," Chinse Journal of Tropical Crops, vol. 30, no. 3, pp. 274–278, 2009.

20.  K. J. Livak and T. D. Schmittgen, "Analysis of relative gene expression data using real-time quantitative PCR and the $2^-$ CT method," Methods, vol. 25, no. 4, pp. 402–408, 2001.

21.  M. Bartolf, E. Brennan, and C. A. Price, "Partial characterization of a cadmium-binding protein from the roots of cadmium-treated tomato," Plant Physiology, vol. 66, no. 3, pp. 438–441, 1980.

22.  J. Zhou and P. B. Goldsbrough, "Structure, organization and expression of the metallothionein gene family in Arabidopsis," Molecular and General Genetics, vol. 248, no. 3, pp. 318–328, 1995.

23.  H. L. Wong, T. Sakamoto, T. Kawasaki, K. Umemura, and K. Shimamoto, "Down-regulation of metallothionein, a reactive oxygen scavenger, by the small GTPase OsRac1 in rice," Plant Physiology, vol. 135, no. 3, pp. 1447–1456, 2004.

24.  A. Figueira, E. A. Kido, and R. S. Almeida, "Identifying sugarcane expressed sequences associated with nutrient transporters and peptide metal chelators," Genetics and Molecular Biology, vol. 24, pp. 207–220, 2001.

25.  W. J. Guo, W. Bundithya, and P. B. Goldsbrough, "Characterization of the Arabidopsis metallothionein gene family: tissue-specific expression and induction during senescence and in response to copper," New Phytologist, vol. 159, no. 2, pp. 369–381, 2003.

26.  W. J. Guo, M. Meetam, and P. B. Goldsbrough, "Examining the specific contributions of individualArabidopsis metallothioneins to copper distribution and metal tolerance," Plant Physiology, vol. 146, no. 4, pp. 1697–1706, 2008.

27.  H. M. Hsieh, W. K. Liu, A. Chang, and P. C. Huang, "RNA expression patterns of a type 2 metallothionein-like gene from rice," Plant Molecular Biology, vol. 32, no. 3, pp. 525–529, 1996.

28.  J. Zhu, Q. Zhang, R. Wu, and Z. Zhang, "HbMT2, an ethephon-induced metallothionein gene fromHevea brasiliensis responds

to $H_2O_2$ stress," Plant Physiology and Biochemistry, vol. 48, no. 8, pp. 710–715, 2010.

29. A. Murphy, J. Zhou, P. B. Goldsbrough, and L. Taiz, "Purification and immunological identification of metallothioneins 1 and 2 from Arabidopsk thaliana," Plant Physiology, vol. 113, no. 4, pp. 1293–1301, 1997.

30. K. Bilecen, U. H. Ozturk, A. D. Duru et al., "Triticum durum metallothionein: isolation of the gene and structural characterization of the protein using solution scattering and molecular modeling," Journal of Biological Chemistry, vol. 280, no. 14, pp. 13701–13711, 2005.

31. W. H. Tang, J. L. Zhang, Z. Y. Wang, and M. M. Hong, "The cause of deviation made in determining the molecular weight of His-tag fusion proteins by SDS-PAGE," Acta Phytophysio-Logica Sinica, vol. 26, no. 1, pp. 64–68, 2000.

32. A. M. Tommey, J. Shi, W. P. Lindsay, P. E. Urwin, and N. J. Robinson, "Expression of the pea gene PsMT(A) in E. coli: metal binding properties of the expressed protein," FEBS Letters, vol. 292, no. 1-2, pp. 48–52, 1991.

33. K. M. Evans, J. A. Gatehouse, W. P. Lindsay, J. Shi, A. M. Tommey, and N. J. Robinson, "Expression of the pea metallothionein-like gene PsMTA in Escherichia coli and Arabidopsis thaliana and analysis of trace metal ion accumulation: implications for PsMTA function," Plant Molecular Biology, vol. 20, no. 6, pp. 1019–1028, 1992.

34. M. García-Hernández, A. Murphy, and L. Taiz, "Metallothioneins 1 and 2 have distinct but overlapping expression patterns in Arabidopsis," Plant Physiology, vol. 118, no. 2, pp. 387–397, 1998.

35. Y. Li and M. A. Trush, "DNA damage resulting from the oxidation of hydroquinone by copper: role for a Cu(II)/Cu(I)) redox cycle and reactive oxygen generation," Carcinogenesis, vol. 14, no. 7, pp. 1303–1311, 1993.

36. Y. Li and M. A. Trush, "Oxidation of hydroquinone by copper: chemical mechanism and biological effects," Archives of Biochemistry and Biophysics, vol. 300, no. 1, pp. 346–355, 1993.

37.  A. Rivetta, N. Negrini, and M. Cocucci, "Involvement of $Ca^{2+}$-calmodulin in $Cd^{2+}$ toxicity during the early phases of radish (Raphanus sativus L.) seed germination," Plant, Cell and Environment, vol. 20, no. 5, pp. 600–608, 1997.

# Chapter 7

# Hydrolysis of Biomass Mediated by Cellulases for the Production of Sugars

Rosa Estela Quiroz-Castañeda[1] and
Jorge Luis Folch-Mallol[1]

[1]Biotechnology Research Centre, Autonomous University of Morelos, Cuernavaca, Morelos, México

## INTRODUCTION

Cellulose, the most abundant organic molecule on Earth is found mainly as a structural component of plant and algal cell walls, is also produced by some animals, such as tunicates, and several bacteria [1]. Natural cellulose is a crystalline and linear polymer of thousands of D-glucose residues linked by β-1, 4-glycosidic bonds, considered the most abundant and renewable biomass resource and a formidable reserve of raw material.

It does not accumulate in the environment due to the existence of cellulolytic fungi and bacteria, which slowly degrade some of the components of plant cell walls. Both fungi and bacteria possess enzymes such as laccases, hemicellulases and cellulases, which efficiently degrade lignin, hemicellulose and cellulose, respectively [2-3].

In plant cell walls the cellulose microfibrils are encrusted in lignin and hemicellulose in a complex architecture that, together with the crystallinity of cellulose, makes untreated cellulosic biomass recalcitrant to hydrolysis to fermentable sugars [4]. However, a group of proteins with cellulose disrupting activity (expansins, expansin-like proteins, swollenins and loosenins) have the capacity of relaxing cell wall tension by disrupting the hydrogen bonds binding together cellulose fibrils and cellulose and other polysaccharides through a non-enzymatic process, improving subsequent sugar releasing [5-8].

An efficient degradation of this polysaccharide content into fermentable sugars could improve the production of biofuels. Rising energy consumption, depletion of fossil fuels and increased environmental concerns have shifted the focus of energy generation towards biofuel use [3].

In this chapter, we focus on cellulose degradation by cellulases in order to enhance sugars release from biomass. Cellulose structure, allomorphs and its hydrolysis by cellulolytic organisms such as fungi and bacteria, is also reviewed, as well as cellulases structure, CAZY classification, their synergistic activity and the recently cellulases identified by metagenomic analysis, an excellent tool in this search of better cellulolytic activity.

Another theme analyzed in this chapter is related to crystalline structure of cellulose, the main impediment to achieve full cellulose hydrolysis, and the role of proteins recently reported with cellulose disrupting activity that have improved saccharification processes. These proteins represent good candidates as an additive to enhance sugar production from plant biomass.

# STRUCTURE AND COMPOSITION OF THE CELL WALL

It has been estimated that the net $CO_2$ fixation by land plants per year is approximately $56 \times 10^9$ tons and that the worldwide biomass production by land plants is $170–200 \times 10^9$ tons (Table 1). Of this amount, 70% is estimated to represent plant cell walls (revised in [9]).

Lignocellulose is a renewable organic material and is the major structural component of all cell plants. Lignocellulose plant biomass consists of three major components: cellulose (40–50 %), hemicellulose (20–40 %) and lignin (20–30 %) (Figure 1).

**Table 1:** Worldwide annual production of biomass

| Production | Tons | Reference |
|---|---|---|
| Assimilated $C_O2$ | 56 X $1^09$ | [10] |
| Plant biomass | 170-200 X $1^09$ | [11] |
| Cell walls | 150-170 X$1^09$ | [9] |
| Lignocellulose | 200 X$1^09$ | [12] |
| Cellulose | 100 X $1^09$ | [13-14] |
| Wheat straw | 540 X $1^09$ | [15] |
| Soybean straw | 200 X $1^09$ | [16] |
| Sugar cane bagasse | 54 X $1^09$ | [17] |

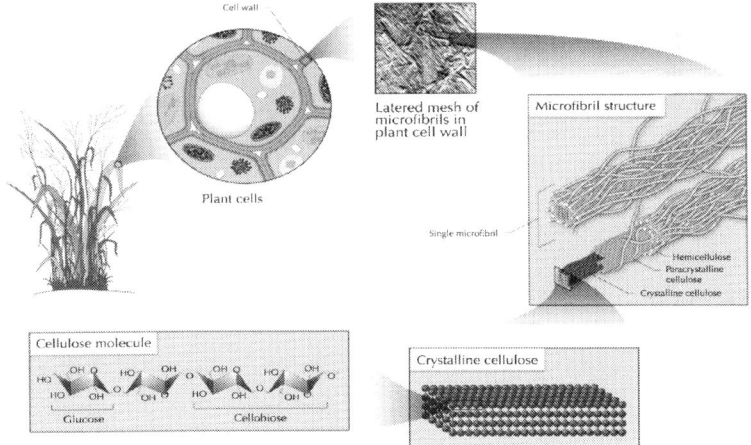

**Figure 1:** Structural organization of the plant cell wall. Cellulose is protected of degradation by hemicelluloses and lignin. Source: Office of Biological and Environmental Research of the U.S. Department of Energy Office of Science. science.energy.gov/ber/.

Minor components are proteins, lipids, pectin, soluble sugars and minerals (Table 2) [9]. It has a thickness of ~0.1 a 10 μm contrasting with <0.01 μm of cell membrane formed by proteins and phospholipids [18].

Examples of such biomass are angiosperms (hardwoods), gymnosperms (softwoods) and graminaceous plants (grasses such as wheat, giant reed and *Miscanthus*).

Cell walls should play a wide array of disparate and sometimes opposing roles: the resistance to mechanical stress is necessary as well as the shape of the cell and protection against pathogens; at the same time, besides it must be reasonably flexible to withstand shear forces, and permeable enough to allow the passage of signalling molecules into the cell [19].

# CELLULOSE STRUCTURE

Cellulose is the main component in the plant cell walls, and is made of parallel unbranched D-glucopyranose units linked by β-1,4-glycosidic bonds that form crystalline and highly organized microfibrils through extensive inter and intramolecular hydrogen bonds and Van der Waals forces, amorphous cellulose correspond to regions where this bonds are broken and the ordered arrangement is lost (Figure 2).The cellulose chains aggregated into microfibrils are reported to consist of 24 to 36 chains based on scattering data and information about the cellulose synthase [20-21].

Consecutive glucose molecules along chains in crystalline cellulose are rotated by 180°, meaning that the disaccharide (cellobiose) is the repeating unit [22].

Two different ending groups are found in each cellulose chain edge. At one end of each of the chains, a non-reducing group is present where a closed ring structure is found. A reducing group with both an aliphatic structure and a carbonyl group is found at the other end of the chains. The cellulose chain is thus a polarized molecule and the new glucose residues are added at the non-reducing end allowing chain elongation (Figure 2) [23].

A wide variety of Gram-positive and Gram-negative bacterial species are reported to produce cellulose, including *Clostridium thermocellum, Streptomyces* spp., *Ruminococcus* spp., *Pseudomonas* spp., *Cellulomonas* spp., *Bacillus* spp., *Serratia, Proteus, Staphylococcus* spp., and *Bacillus subtilis* [24].

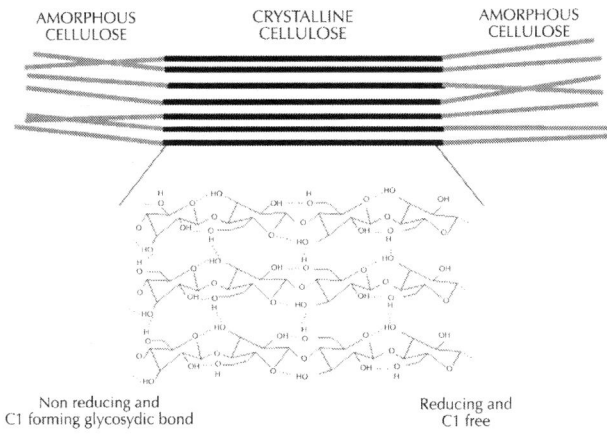

**Figure 2:** Crystalline and amorphous structure of cellulose. The crystalline structure is conserved by hydrogen bonds and Van der Waals forces, in amorphous structure exists twists and torsions that alter the ordered arrangement. Reducing and non-reducing are shown.

# CELLULOSE CRYSTALLINITY

In plants, cellulose is synthesized by CESA proteins (Cellulose Synthase) embedded in plasmatic membrane arranged in hexameric groups called rosettes particles [25].

Cellulose crystallites are thought to be imperfect, the traditional two-phase cellulose model describes cellulose chains as containing both crystalline (ordered) and amorphous (less ordered) regions. Crystalline structure of cellulose implies a structural arrangement in which all atoms are fixed in discrete position with respect to one another. An important feature of the crystalline array is that the component molecules of individual microfibrils are packed sufficiently tightly to prevent penetration not only by enzymes, but even by small molecules such as water. While its recalcitrance to enzymatic degradation may pose problems, one big advantage of cellulose is its homogeneity [1, 26-27].

Highly ordered, crystalline regions are interspersed with regions containing disorganized or amorphous cellulose, which constitute 5 to 20% of the microfibril. Many studies have shown that completely disordered or amorphous cellulose is hydrolysed at a much faster rate than partially crystalline cellulose; this fact supports the idea that the initial degree of crystallinity is important in determining the enzymatic digestibility of a cellulose sample. Crystallinity, is a measure of the weight fraction of the crystalline regions, is one of the most important measurable properties of cellulose that influences its enzymatic digestibility [19, 28-30].

A parameter termed the crystallinity index (CI) has been used to describe the relative amount of crystalline material in cellulose. Generally, in nature, crystallinity indexes range from 40% to 95%, the rest is amorphous cellulose [31]. The degree of polymerization, (DP) is the number of monomeric units in a polymer molecule, which in cellulose it ranges from 500 to 15,000 but varies depending the substrate (Table 2).

**Table 2:** Some physical properties of cellulosic substrates

| Substrate | Crystallinity index | Degree of polymerization | Ref. |
|---|---|---|---|
| Carboxymethyl cellulose (CMC)a | NA | 100-2000 | [32] |
| Cellodextrins a | NA | 2-6 | [32] |
| Avicel b | 0.5-0.6 | 300 | [13] |
| BC b | 0.76-0.95 | 2000 | [13] |
| PASC b | 0-0.04 | 100 | [13] |
| Cotton b | 0.81-0.95 | 1000-3000 | [13] |
| Filter paper b | 0-0.45 | 750 | [13] |
| Wood pulp b | 0.5-0.7 | 500-1500 | [13] |
| Fluka Avicel PH-10 b | 0.56-0.91 | 200-240* | [26] |
| Fluka cellulose b | 0.48-0-82 | 280* | [26] |
| Sigma $\alpha$-cellulose b | 0.64 | 2140-2420* | [33] |

[i]-*According to manufacturer's data. a, Soluble; b, Insoluble.

# CELLULOSE ALLOMORPHS

The crystalline structure of cellulose has been studied since its discovery in the 19th century, its structure was first established by Carl von Nageli in 1858, and the result was later verified by X-ray crystallography [34-35].

In the past decades, many data on the polymorphism of cellulose were analysed, being the most reliable data published after 1984, when the results of NMR spectroscopic studies of cellulose were reported [36].

The repeating unit of the cellulose macromolecule includes six hydroxy groups and three oxygen atoms. Therefore, the presence of six hydrogen bond donors and nine hydrogen bond acceptors provides several possibilities for forming hydrogen bonds. Due to different arrangements of the pyranose rings and the possible conformational changes of the hydroxymethyl groups, cellulose chains can exhibit different crystal packings [37].

Four different crystalline allomorphs of cellulose have been identified by their characteristic X-ray diffraction patterns and solid-state $^{13}C$ nuclear magnetic resonance (NMR) spectra: celluloses I, II, III ($III_I$, $III_{II}$) and IV ($IV_I$, $IV_{II}$). The most important allomorphs are cellulose I and II [22].

Some difference in symmetry and chain geometry have been found in unit cell dimensions of various allomorphs and some parameters have been established: a, interchain distance, b unit chain length and c, intersheet distance, as well as the angles α, β and γ which are the angles between b and c, a and c, and a and b, respectively, (Table 3) [38-40].

**Table 3:** Unit cell parameters of different cellulose allomorphs obtained by X-ray diffractions

| Allomorph | Unit cell parameters | | | | | | Ref. |
|---|---|---|---|---|---|---|---|
| | Bond lengths (Å) | | | Angles (°) | | | |
| | a | b | c | A | β | γ | |
| I | 6.717(7) | 5.962(6) | 10.400(6) | 118.08(5) | 114.80(5) | 80.37(5) | (41) |
| *I* | 7.784(8) | 8.201(8) | 10.380(10) | 90 | 90 | 96.5 | (42) |
| II | 8.10(1) | 9.03(1) | 10.31(1) | 90 | 90 | 117.10(5) | (43) |
| II | 8.03(1) | 9.04(1) | 10.35(1) | 90 | 90 | 117.11(2) | (44) |
| | *8.03(1)(* | *9.02(1)* | *10.34(1)* | 90 | 90 | *117.11(2)* | |
| II₁I | 4.450(4) | 7.850(8) | 10.310(10) | 90 | 90 | 105.10(5) | (45) |

[i] - I, Fresh water algae *Glaucocystis nostochinearum;* I$_\beta$, Tunicate *Halocynthia roretzi;* II, Ramie cellulose (mercerized); II, Regenerated cellulose (Fortisan); III, Marine algae *Cladophora*. All crystal structures have been determined at 293°K, except allomorph II (Fortisan) that was also determined at 100°K (italics).

*Cellulose I* is the most abundant form found in nature, is a mixture of two distinct crystalline forms: cellulose I$_\alpha$, the predominant form isolated from bacteria (*Acetobacter xylinum*) and fresh water algae (*Glaucosystis nostochinearum*); and cellulose I$_\beta$ is the major form in higher plants such as cotton and wood celluloses, ramie and animal celluloses, for example in the edible ascidian *Halocynthia roretzi*[4]. Cellulose from the marine algae *Claudophora* sp. and *Valonia ventricosa* is a mixture of both forms, predominating I$_\alpha$ [37]. Currently, cellulose I is receiving increased attention due to its potential use in bioenergy production.

*Cellulose I$_\alpha$* has a triclinic one-chain unit cell where parallel cellulose chains stack through van der Waals interactions, with progressive shear parallel to the chain axis. *Cellulose I$_\beta$* has a monoclinic two-chain unit cell, which means parallel cellulose chains stacked with alternating shear (Figure 3) [46].

*Cellulose II* is the most crystalline thermodynamic stable form, it can also be obtained from cellulose I by two distinct routes: mercerization (alkali treatment) and regeneration (solubilization and subsequent recrystallization) [47]. Cellulose II, like cellulose I$_\beta$, has the monoclinic unit cell (space group P2$_1$). The different arrangement of the chains (parallel in cellulose I$_\beta$ and antiparallel in cellulose II) is the most substantial difference between these two polymorphs. The cellulose is a highly rigid macromolecule due to the presence of a three-dimensional hydrogen bond network in addition to the C-O-C bonds between the glucopyranose rings. In the absence of such hydrogen bond networks the chains are much more flexible. These hydrogen bonds are responsible for both the poor solubility of cellulose and the difference in the reactivity of the hydroxy groups in esterification reactions (Figure 4) [37].

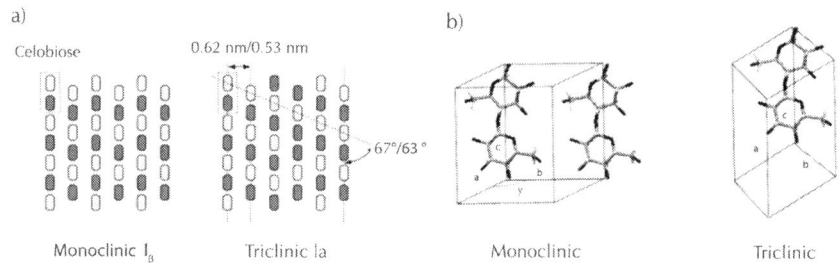

a)

Celobiose    0.62 nm/0.53 nm

Monoclinic $I_\beta$      Triclinic Ia

b)

Monoclinic      Triclinic

67°/63 °

**Figure 3:** Differences between the monoclinic and triclinic forms of cellulose I. a) In the monoclinic form, cellobiose units stagger with a shift of a quarter of the c-axis period (0.26 nm), whereas the triclinic form exhibits a diagonal shift of the same amount. The angles shown depend on which crystallographic face is being viewed. A glucose unit is represented by rectangles (cellobiose, a dimer of glucose); image reproduced with publisher´s permission [23]. b) Mode of packing in the unit cell of cellulose I: mono and triclinic unit cell. Notice that the monoclinic angle γ is obtuse. Image reproduced with permission from PNAS Copyright (2012).

*Cellulose III,* and *III,,* can be formed from cellulose I and II, respectively, by treatment with ammonia; in a reversible reaction. Besides producing the different allomorphs of cellulose, this chemical treatment can also alter other physical properties of cellulose, such as the degree of crystallinity and therefore enhanced cellulase accessibility and chemical reactivity. The degree of conversion of cellulose I to cellulose IIII depends on the reaction period and the temperature used in the final stage of the treatment [47-48].

In [45] solved the crystal structure of cellulose IIII by synchrotron X-ray and neutron fiber diffraction analyses, and showed that it has a lower packing density than cellulose $I_\alpha$ or $I_\beta$ (Figure 4).

*Cellulose IV* can be most easily prepared by heating cellulose III, and therefore, two polymorphs of it also exist -celluloses $IV_I$ and $IV_{II}$ obtained respectively, from celluloses $III_I$ and $III_{II.}$ In general, cellulose IV could be prepared by treatment in glycerol at 260 °C after transformation into cellulose II or III. Cellulose I cannot be transformed directly into cellulose IV [46, 49].

Fibrillation makes cellulose $IV_I$ less suitable for crystallographic analysis: that is, it makes it more difficult to interpret cellulose $IV_I$ as a crystal. For these reasons, it is unclear whether is a crystal with an orthogonal unit cell or a less crystalline form of cellulose I [49]. A thorough review of cellulose crystalline allomorphs can be found elsewhere [46-47].

Although considerable progress has been made in elucidating the crystal structures of cellulose in microfibrils, they are still not well understood, and a deeper understanding of cellulose structure is required [50-51].

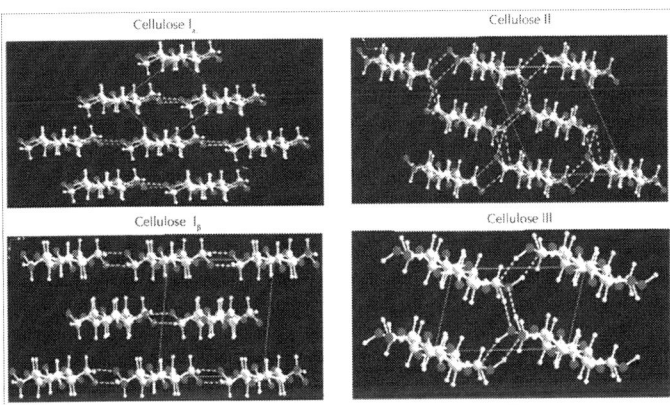

**Figure 4:** Projections of the crystal structures of cellulose I (α,and β) II and III down the chain axes directions. C, O, and H atoms are represented as gray, red, and white balls, respectively. Covalent and hydrogen bonds are represented as full and dashed sticks, respectively. H atoms involved in hydrogen bonding are explicitly represented for only cellulose IIII. Only the major components of hydrogen bonds are represented. Adapted with permission from [45]. Copyright (2012) American Chemical Society.

# CELLULOSE-DEGRADING MICROORGANISMS

Since cellulose is very difficult to degrade as a component of plant cell walls, only a few microorganisms specialized for plant cell wall degradation can hydrolyse cellulose. Among these, anaerobic and aerobic genera of Domain Bacteria and fungi of Domain Eukarya are included.

Generally speaking, two types of systems occur in regards to plant cell wall degradation by microorganisms. In one type, the organism produces a set of free enzymes that act synergistically to degrade plant cell walls. In the second type, the degradative enzymes are organized into an enzyme complex located in cellular surface called the cellulosome. This complex is very effective in degrading plant cell walls [52].

Anaerobic and aerobic bacteria have different strategies to degrade cellulolytic substrates; whereas anaerobic bacteria degrade cellulose using cellulosomes, aerobic bacteria secretes enzymes capable of degrading cellulose that freely diffuse to reach the substrate.

Anaerobic bacteria of the order *Clostridiales* (Phylum *Firmicutes*) are generally found in soils, decaying plant waste, the rumen of ruminant animals, compost, waste water, and wood processing plants; these bacteria have also been found in insects like termites (*Isopteran*), bookworm (*Lepidoptera*), and so, in a symbiotic relationship in their guts responsible for cellulosic feed digestion. Anaerobic hydrolysis represents 5% to 10% of global cellulose degradation [53-55].

Aerobic bacteria with cellulolytic activities of the order *Actinomycetales* (phylum *Actinobacteria*) have been found on soils, water, humus, agricultural waste (sugar cane) and decaying leaves, these bacteria excretes enzymes capable of degrading cellulose (cellulases) [52]. In aerobic bacteria *Pseudomonas fluorescens* subsp. cellulosa, *Streptomyces lividans* and *Cellulomonas fimi* cellulolytic systems of degradation have been reported [56-58].

Some anaerobic bacteria with cellulolytic activity are Butyrivibrio fibrisolvens, Fibrobacter succinogenes, Ruminococcus flavefaciens, Clostridium cellulovorans, C. cellulolyticum and C. thermocellum [59-61].

Due to the significant diversity in the physiology of cellulolytic bacteria, sometimes is difficult to classify bacteria as mentioned above, therefore, on this basis, they can be placed into three diverse physiological groups: (1) fermentative anaerobes, typically Gram-positive, (*Clostridium* and*Ruminococcus*), but with a few Gram-negative species (*Butyvibrio* and *Acetivibrio*) that are phylogenetically related to the *Clostridium* assemblage (*Fibrobacter*); (2) aerobic Gram-positive bacteria (*Cellulomonas* and *Thermobifida*) and (3) aerobic gliding bacteria, (*Cytophaga* and*Sporocytophaga*) [1, 53].

The ability to utilize lignocellulosic material is widely distributed among fungi, from chytridiomycetes to basidiomycetes. Among fungi, the most efficient at using wood as substrate are the basidiomycetes, considered the principal taxonomic group involved in the aerobic degradation of wood with all its components, they are the main organic material decomposition agents. These aerobic fungi produce extracellular enzymes allowing lignocellulose degradation (lacasses, hemicellulases, and cellulases), although some Ascomycetes are able to degrade cellulosic compounds as well. Unlike aerobic fungi, some of the Chytridiomycetes anaerobic fungi, have multienzymatic complexes similar to cellulosomes of bacteria [1, 3, 62-63] some members are anaerobic species living in the gastrointestinal tract of ruminants such as *Anaeromyces, Caecomyces, Neocallimastix, Orpinomyces* and *Piromyces.*

Examining the taxonomic composition of cellulolytic fungi inhabiting the decaying leaves and rotting woods of forest soils, zygomycetes are represented by a single genus, *Mucor,* while ascomycetes and basidiomycetes are represented by genera such as *Chaetomium, Trichoderma, Aspergillus, Penicillium,Fusarium, Coriolus, Phanerochaete, Schizophyllum, Volvariella, Pycnoporus* and *Bjerkandera.* Two of the most studied fungi, due to their industrial relevance, are *Trichoderma reesei* and *Phanerochaetechrysosporium.*

Nowadays, more than 14,000 fungi, which are active against cellulose and other insoluble fibres, are known [1, 24, 64-66]. A more detailed list of cellulose degrading bacteria and fungi is listed in Table 5.

# CELLULOSE DEGRADATION MEDIATED BY CELLULOSOME

Selective pressure of evolution is the force driving microorganisms to adapt a new environment, in anaerobic conditions is necessary a machinery for the extracellular degradation of substrates, such as the recalcitrant crystalline components of the plant cell wall. Due to this, the anaerobes tend to adopt different strategies for degrading plant components, being the cellulosomes the most remarkable feature.

**Table 4:** Fungi and bacteria with cellulolytic activity

| Group | | Fungi | Enzymes** | Substrate | Ref |
|---|---|---|---|---|---|
| Aerobic fungi (extracellular cellulolytic enzymes) | Ascomycetes | T. reesei | Cel, Xyl | Wheat straw | [68-69] |
| | | T. harzianum | Cel | Cellulose | [70] |
| | | A. niger | Cel | Sugar cane bagasse | [71] |
| | | Schizophyllum commune | Cel, Xyl | Microcrystalline cellulose, rice xylan | [72] |
| | Basidiomycetes | P. chrysosporium | Cel, Xyl | Red oak, grape seed, barley bran, sorghum | [2, 63, 73] |
| | | B. adusta Pycnoporus sanguineus | Cel, Xyl | Oak and cedar sawdust, rice husk, corn stubble, wheat straw and Jatrophaseed husk | [74] |
| | | Fomitopsis palustris | Cel | Microcrystalline cellulose | [75] |
| Anaerobic rumen fungi (cellulosomes) | Chytridiomycetes | Anaeromyces mucrunatus | Cel, Xyl | Orchard grass hay | [76] |
| | | Caecomyces communis | Cel | Microcrystalline cellulose, alfalfa hay | [77] |
| | | Neocallimastix frontalis | Cel, Xyl | wheat straw | [78] |
| | | Orpinomyces sp. | Cel | Avicel | [79] |
| | | N. patriciarum | Cel | CMC | [80] |

| Aerobic bacteria (cellulosomes) | Actinobacteria | Acidothermus cellulolyticus | Cel | Whatman paper No1, Microcyrtsalline cellulose, | [81] |
| | | Actinospica robiniae | | Azurine crosslinked hydroxyethylcellulose (AZCL-HEC) | |
| | | Actinosynnema mirum | | | |
| | | Catenulispora acidiphila | | | |
| | | Cellulomonas flavigena | | | |
| | | Thermobispora bispora | | | |
| | | Xylanimonas cellulosilytica | | | |
| Anaerobic bacteria (cellulosomes) | Firmicutes | Clostridium thermocellum* | Cel | Crystalline cellulose | [82] |
| | | Thermomonospora fusca* | Cel, Xyl | wheat straw, oat spelt xylan | [83] |
| | | Caldicellulosiruptor kristjanssonii* | Cel | Microcrystalline cellulose | [84] |
| | | Anaerocellum thermophilum* | Cel, Xyl | Microcrystalline cellulose, xylan | [85-86] |

[i] - *Termophylic bacteria. **Cel: cellulases; Xyl: xylanases.

The occurrence of a cellulosome was first observed in the thermophilic bacterium, *C. thermocellum*and has now been described in a number of mesophilic anaerobic bacteria and with some anaerobic fungi particularly *Piromyces* sp.[52, 67].

Cellulosomes are large extracellular enzyme complexes capable of degrading cellulose, hemicelluloses, and pectin; they may be the largest extracellular enzyme complexes found in nature, although the individual cellulosomes size range from 0.65 MDa to 2.5 MDa, some polycellulosomes have been reported to be as large as 100 MDa, [87].

The cellulosome structure is characterized by two components: (a) the non-enzymatic scaffolding proteins with enzyme binding sites called cohesins, (b) enzymes with dockerins proteind interacting with cohesins in the scaffolding protein (Figure 5).

Depending of the bacterial species, the scaffolding protein varies in the number of cohesins and cellulose binding modules (CBM) that binds the cellulosome tightly to the substrate and concentrates the enzymes to a particular site of the substrate. Recently, a more complex cellulosome structure with multiple interacting scaffolding proteins that allows the binding of as much as enzymes has been revealed [52].

The cohesin-dockerin interconnect the different scaffoldin components, whereby the specificities among the individual cohesin-dockerin complexes dictate the overall supramolecular architecture of the participating components [88].

In short, the enzymatic cellulosome system may exceed the potential of non-cellulosomal degradative system due to its structural organization, efficient binding to the substrate, the variety of hydrolytic enzymes acting synergistically [52]. Cellulosomes have not been identified in bacteria (or eukarya) that grow above 65 °C, and have not been identified in the Archaea [89].

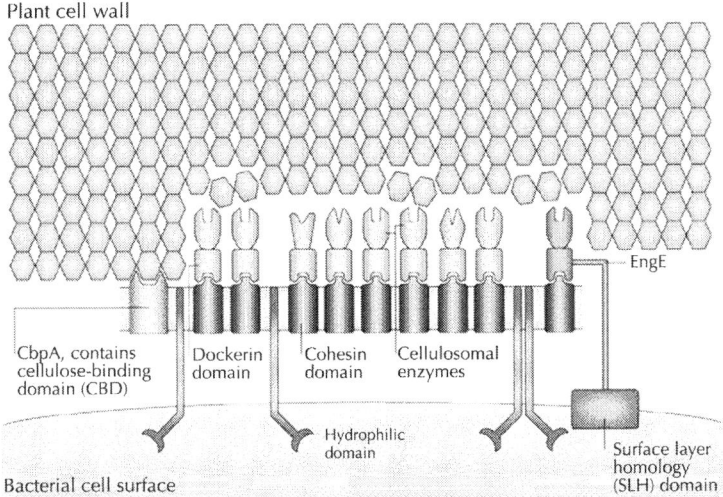

**Figure 5:** Cellulosome structure. A dockerin is appended to catalytic (enzyme) and noncatalytic carbohydrate-binding modules (CBMs). Dockerins bind the cohesins of a noncatalytic scaffoldin, providing a mechanism for cellulosome assembly. Image reproduced with publisher´s permission [90].

# CELLULOSE DEGRADATION MEDIATED BY NON-CELLULOSOMAL ENZYMES

Aerobic cellulolytic bacteria and fungi use a system for cellulose degradation consisting of sets of soluble cellulases. Cellulases are inducible enzymes by cellulosic substrates, which are synthesized by a large diversity of microorganisms including both fungi and bacteria during their growth on cellulosic materials.These microorganisms can be aerobic, anaerobic, mesophilic or thermophilic. Among them, the genera of *Clostridium, Cellulomonas, Thermomonospora, Trichoderma,* and *Aspergillus* are the most extensively studied cellulases producers [91].

The aerobic cellulase mechanism evolved in terrestrial microorganisms that colonise solid substrates and therefore secrete cellulases to enable degradation of the substrate. Because of the recalcitrance of plant cell walls some cellulolytic microorganisms secrete up to 50% of their total protein during growth on biomass or cellulose [53, 90].

Cellulases are composed of independently folding and structurally and functionally discrete units called domains, making cellulases modular enzymes. Structurally fungal cellulases are simpler as compared to bacterial cellulosomes [32, 88, 92].

Fungal cellulases have two independent domains: a catalytic domain (CD) and a cellulose-binding domain (CBD), which is joined by a short poly linker region to the catalytic domain at the N-terminal. The CBD is comprised of approximately 35 amino acids, and the linker region is rich in serine and threonine [93].

It is clear that the role of the CBD is to bind the enzyme to the cellulose so that the CD keep closer to the substrate and it also gives the CD time to move the chain into its active site before the enzyme diffuses away from the cellulose. It is still not clear whether the CBD also can modify cellulose or otherwise assist cellulose hydrolysis by the catalytic domain [94].

The mixture of free cellulases act synergistically to degrade crystalline cellulose increasing the specific activity up to fifteen fold higher than that of any individual cellulase [95].

# CELLULOLYTIC ORGANISMS FROM EXTREME ENVIRONMENTS

Novel enzymes with application in industry require improved features to tolerate extreme conditions of temperature, pH and salinity. Some microorganisms live in these environments, so called extremophiles and are considered a source of enzymes with potential biotechnological applications.

Extreme environments host a number of cellulolytic microorganisms, such as the Gram–negative Antarctic bacterium *Pseudoalteromonas haloplanktis,* collected from seawater, which secretes a psychrophilic cellulase, Cel5G, this cold adapted enzyme displays a high specific activity at low and moderate temperatures and a rather high thermosensitivity induced by a decrease of the intramolecular interactions [96-97].

Extremely thermophilic cellulose-degrading microorganisms are of particular and biotechnological interest owing to the presence of highly thermostable enzymes. A deeper analysis of these organisms is reported in [89, 98-99].

The group of thermophilic cellulolytic prokaryotes includes two aerobic species, *Rhodothermus marinus* and *Acidothermus cellulolyticus*, and numerous anaerobes of the genera *Caldicellulosir uptor*,*Clostridium*, *Spirochaeta*, *Fervidobacterium* and *Thermotoga* (reviewed by (100)).

All members of the genus *Caldicellulosiruptor* are extremely thermophilic, cellulolytic, and non-spore-forming anaerobes with Gram-positive type cell wall, capable of fermenting different types of carbohydrates and have been isolated mostly from neutral or slightly alkaline geothermal springs in New Zealand, Iceland and California [100].

Recently, thermostable cellulases have also been reported in the thermophilic *Geobacillus* sp. R7 that produces a cellulase with a high hydrolytic potential when grown on pretreated agricultural residues (corn stover and prairie cord grass). In fact, it was demonstrated that *Geobacillus* sp. R7 can ferment the lignocellulosic substrates to ethanol in a single step, improving bioethanol production with important potential for cost reductions. Cellulases genes were also identified in several *Sulfolobales*strains, however, their physiological function is not well understood [101-102].

Another thermophlic bacterium *Anaerocellum thermophilum* degrade lignocellulosic biomass untreated as well as crystalline cellulose and xylan [86].

While cellulases are widespread in Fungi and Bacteria, only one archaeal cellulase, an endoglucanase from *Pyrococcus furiosus*, has been reported. This enzyme exhibits a significant hydrolyzing activity toward crystalline cellulose even though it lacks a CBD, the role of this intracellular enzyme in Archaea is unclear, given that Archaea are apparently unable to grow on cellulose [103-104].

In the alkali tolerant fungus *Penicillium citrinum* an alkali tolerant and thermostable cellulases were found which may have potential effectiveness as additives to laundry detergents [105].

In this search to improve cellulases activity, hybrids of hyperthermostable glycoside hydrolases have been constructed as reported by [106], for example, using the structural compatibility of two hyperthermostable family 1 glycoside hydrolases, *P. furiosus* CelB and *Sulfolobus solfataricus* LacS a library of hybrids using DNA family shuffling was created.

This study demonstrates that extremely thermostable enzymes with limited homology and different mechanisms of stabilization can be efficiently shuffled to form stable hybrids with improved catalytic features.

Alkaliphilic, thermophylic and halophilic microbial species have the potential to yield valuable new products for biotechnological industry. Alkaliphilic polymer-degrading enzymes such as proteases, lipases and cellulases are most frequently isolated from *Bacillus* or related species. Cellulases and lipases are important not only as components of washing detergents, but they are also applied in the paper and pulp, pharmaceutical, food, leather, chemical or waste treatment industries [107-108].

# CELLULASES STRUCTURE

Proteins with hydrolytic activity such as cellulases and hemicellulases comprises a complex molecular architecture of discrete modules (a catalytic domain (CD) and one or more CBDs), which are joined by unstructured linker sequences [109].

The *catalytic domain* spans more than 70% of protein sequence. A sequence analysis of these domains in different cellulases shows a significant variability between them, in fact, active site of the enzyme has distinct three dimensional arrangements: in tunnel shape for a processive exo degradation or in a cleft shape for an endo degradation. This domain is N-glycosylated and is responsible of the cleavage of the glycosidic bond, which occurs through an acid hydrolysis mechanism, using a donor of protons and a nucleophyle or base such as glutamic and aspartic acid [1, 110-111].

The *cellulose binding domain* facilitates hydrolysis by keeping the catalytic domain nearby the substrate, therefore the presence of CBD is important for cellulases starting and processivity [112]. The CBDs, which is usually O-glycosylated, contain from 30 to about 200 amino acids, and exist as a single, double, or triple domain in a protein. Their location in the protein can be both, C or N terminal and occasionally is centrally positioned.

The CBDs bring the enzyme into a closer and prolonged association with the substrate, increasing the rate of catalysis, this domain was found to function more efficiently in substrate degradation, and removing the CBM from the enzyme or from the scaffolding in cellulosomes dramatically decrease its enzymatic activity (revised in [109]).

In the union of CBD and cellulose, some non-polar residues left exposed, mostly tyrosines and tryptophans, showing the flat face of their aromatic ring towards the pyranose ring, this interaction is stabilized by polar residues that form hydrogen bonds [61].

Besides cellulases, CBDs have also been found in other polysaccharides degrading enzymes: hemicellulases, endomannanases and acetilxylanesterases [113].

The *linker peptide* is a sequence of amino acids connecting the cellulose binding domain and the catalytic domain. This linker contains from 6 to 59 amino acids and functions as a flexible hinge that allows the independent function of each domain [114]. The sequence of the linker varies between enzymes, however, the composition is typically rich in proline, treonine and serine, like in the sequence $PTPTPTPTT(PT)_7$ of the endoglucanase of *C. fimi* and NPSGGNPPGGNPPGTTTTRRPATTTGSSPG of the cellobiohydrolase CBHI of *T. reesei*.

Treonine and serine residues of the peptide linker are highly O-glycosylated to be protected from proteolysis; if the linker is completely absent or is too short then both domains, CBD and CD, obstruct each other and the affinity reduces. Based on the similarities of the linker between cellulases it has been suggested that it could be acting as a flexible hinge facilitating independent function of the domains (Figure 6) [115-116].

# MECHANISMS OF CELLULOSE BIO-DEGRADATION

Once the cellulase has recognized a free chain end, it threads the chain into the tunnel to form a catalytically active complex (CAC). Because cellulose decrystallization in water is free-energetically unfavourable, the tunnels or clefts of cellulase CDs contain hydrophobic and polar residues that form favourable contacts with the chain.

Several studies have mutated hydrophobic residues in the CD tunnels of cellulases and chitinases (structurally similar to cellulases), and have demonstrated that hydrophobic residues need to be present in the CD tunnels for digestion of crystalline cellulose to occur [117].

Additionally, in [27] have shown that removal of hydrophobic residues in cellulase and chitinase tunnels can increase processivity rates on more accessible polymers.

Once a cellulase forms a CAC with a cellodextrin chain, the hydrolysis reaction occurs usually via a retaining or inverting mechanism, depending on the directionality of the enzyme. After the reaction occurs, the product must be expelled and another CAC formed by threading another cellobiose unit into the CD (Figure 6) [117].

In most cases, the hydrolysis of the glycosidic bond is catalysed by two amino acid residues of the enzyme: a general acid (proton donor) and a nucleophile/base [111]. Depending on the spatial position of these catalytic residues, hydrolysis occurs via overall retention or overall inversion of the anomeric carbon. Recently, a completely unrelated mechanism has been demonstrated for two families of glycosidases utilizing $NAD^+$ as a cofactor [118-119].

The *retaining glycoside hydrolase mechanism* leads to a net retention of the configuration at the anomeric carbon (C1) of the substrate after cleavage, since the hydrolysis of a glycosidic bond creates a product with the same configuration at the anomeric carbon as the substrate had before hydrolysis.

The *inverting glycoside hydrolase mechanism* leads to a net inversion of the configuration at the anomeric carbon (C1) of the substrate after cleavage. This is performed via a single nucleophilic displacement mechanism, where the hydrolysis of a -glycosidic bond creates a product with the -configuration, and vice-versa [120].

# CELLULOSE BIODEGRADATION

Although more than a dozen fungal species considered as cellulose degraders have been reported (including *T. viride, T. reesei, F. solani, A. niger, A. terreus, P. chrysosporium, B. adusta* and *P. sanguineus)* [3, 74];

and even with cellulases identified in nematodes (*Bursaphelenchus xylophilus,* a nematode infecting pine wood), yeast (*Aureobasidium pullulans*) and marine bacteria (*Saccharophagus degradans*), the search of new cellulases genes continues. This have led to the construction of metagenomic libraries and bioprospecting analysis from several environments: buffalo rumen, higher termite guts, bovine ruminal protozoan, decomposing poplar wood chips and hardwood forest leading to the identification of new genes and organisms with cellulolytic activities [107, 121-130].

To have a better impression of the latest developments regarding fungal carbohydrate-active enzymes, the following sections will discuss the enzymes needed for cellulose degradation.

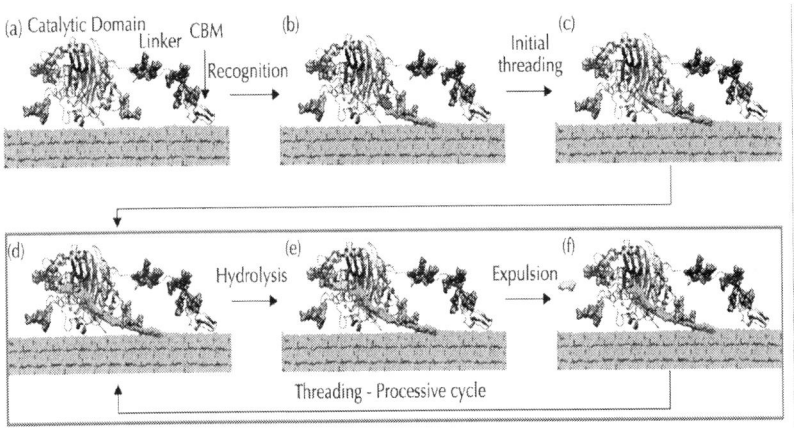

**Figure 6:** Activity on substrate of cellulase (exoglucanase, Cel7A) of *T. reesei.* The enzyme has a small carbohydrate-binding domain (CBD) of 36-amino acid, a long flexible linker with O-glycan (dark blue), and a large catalytic domain (CD) with N-linked glycan (pink) that can thread a single chain of cellulose into the catalytic tunnel of 50 Å.

Cel7A binding to cellulose, (*b*) recognition of a reducing end of a cellulose chain, (*c*) initial threading of the cellulose chain into the catalytic tunnel, (*d* ) threading and formation of a catalytically active complex, (*e*) hydrolysis in a processive cycle and ( *f* ) product expulsion and threading of another cellobiose (shown in yellow in *e* and *f*). Image reproduced with publisher´s permission [131].

# CELLULASES

Multiple types of modular cellulases formed by catalytic and carbohydrate binding domains have been discovered, including at least two exo-β-glucanases, or cellobiohydrolases (CBHs,CBH I and CBH II), four endoglucanases (EG; EG I, EG II, EG III, EG V), and one β-glucosidase (BG) [1].

Cellulases are O-glucoside hydrolases (GH, EC 3.2.1.), a widespread group of enzymes which hydrolyse the β-1,4 linkages or glycosidic bond between two or more carbohydrates or between a carbohydrate and a non-carbohydrate moiety. GH are classified into cellulases families on the basis of amino acid sequence similarity [31, 132].

A classification of glycoside hydrolases in families based on amino acid sequence similarities has been proposed a few years ago. Because there is a direct relationship between sequence and folding similarities, this classification reflects the structural features of these enzymes better than their sole substrate specificity, and helps to reveal the evolutionary relationships between these enzymes, which represent a convenient tool to deduce information of the mechanism [132-133].

Out of the currently existing 125 families, 15 correspond to cellulases (GHF 1,3, 5, 6, 7, 8, 9, 12, 44, 45, 48, 51, 74, 116, and 124), and 64 families group the cellulose binding domains (see http://www.cazy.org/). In [134] an excellent review of and classification system for many CBD families is provided.

The widely accepted mechanism for enzymatic cellulose hydrolysis involves synergistic actions by endoglucanases (EGL, EC 3.2.1.4], exoglucanases or cellobiohydrolases (CBH, EC 3.2.1.74; 1,4-β-D-glucan-glucanhydrolase and EC 3.2.1.91; 1,4-β-D-glucan cellobiohydrolase), and β-glucosidases (BGL, EC 3.2.1.21).

Endoglucanases hydrolyse accessible intramolecular β-1,4-glucosidic bonds of cellulose chains randomly to produce new chain ends; exoglucanases processively cleave cellulose chains at the reducing and non-reducing ends to release soluble cellobiose or glucose; and β-glucosidases hydrolyse cellobiose to glucose in order to eliminate cellobiose inhibition (13). These three hydrolysis processes occur simultaneously as shown in Figure 7.

The activity of cellulase enzyme systems is much higher than the sum of the activity of its individual subunits; a phenomenon known as synergism, so they have to be considered not just simply a conglomerate of enzymes with components from all three cellulase types, but as a mixture that efficiently hydrolyse cellulose fibres.

# ENDOGLUCANASES

These enzymes cleave internal linkages in amorphous cellulose filaments, generating oligosaccharides with different sizes and creating new chain ends that can in turn be attacked by exoglucanases (135). The cellulolytic process is initiated by endoglucanases that randomly cleave internal linkages at amorphous regions of the cellulose fibre and creating new reducing and non-reducing ends that are susceptible to the action of cellobiohydrolases [136].

Endoglucanases are monomeric enzymes with a molecular weight that ranges from 22 to 45 kDa, although some fungi such as *Sclerotium rolfsii* and *Gloeophyllum sepiarium* have endoglucanases twice this size [137]. In general, endoglucanases are not glycosylated; however, they sometimes may have relatively low amounts of carbohydrate (from 1 to 12%) [2]. Unlike other endoglucanases reported with optimum pH 4 to 5; the only known endoglucanase with a neutral pH optimum is that from the basidiomycete *Volvariella volvacea,* expressed in recombinant yeast. Basically, their optimum temperature ranges from 50 to 70 °C [138-139].

Exhaustively hydrolysing cellulose also requires the action of β-glucosidases (BGL) (EC 3.2.1.21), which hydrolyse cellobiose, releasing two molecules of glucose and thereby provide a carbon source that is easy to metabolize. Fungi causing white and brown rot, mycorrhizal fungi and plant pathogens produce these enzymes [2, 135].

According to [13], primary hydrolysis occurs on the surface of solid substrates and releases soluble sugars with a degree of polymerization (DP) up to 6 into the liquid phase upon hydrolysis by endoglucanases and exoglucanases. This depolymerisation step performed by endoglucanases and exoglucanases is the rate-limiting step for the whole cellulose hydrolysis process. The second hydrolysis involves primarily the hydrolysis of cellobiose to glucose by β-glucosidases, although some β-glucosidases also hydrolyse longer cellodextrins. The combined actions of endoglucanases and exoglucanases modify the cellulose surface characteristics over time, resulting in rapid changes in hydrolysis rates [32].

To assay endoglucanase activity, there are substrates that are used, such as carboxymethylcellulose (CMC), a soluble amorphous cellulose form that is an excellent substrate for endocellulases and its hydrolysis does not require a CBD [110].

# EXOGLUCANASES

Also known as cellobiohydrolases, these enzymes catalyse the successive hydrolysis of residues from the reducing and non-reducing ends of the cellulose, releasing cellobiose molecules as main product, which are hydrolysed by β-glucosidases. They account for 40 to 70% of the total component of the cellulase system, and are able to hydrolyse crystalline cellulose.

Exoglucanases have shown specificity on the ends of cellulose, such as *T. reesei* cellobiohydrolase (CBH) I and II that act on the reducing and non-reducing cellulose chain ends, respectively [112].

These enzymes are monomeric proteins with a molecular weight ranging from 50 to 65 kDa, although there are smaller variants (41.5 kDa) in some fungi, such as *Sclerotium rolfsii*. Low levels of glycosylation (around 12% to none at all) are found in these enzymes; and their optimum pH is 4 to 5, with an optimum temperature from 37 to 60 °C, depending on the specific enzyme-substrate combination [137, 140].

Exoglucanases form part of the cellulolytic machinery of the fungi causing white and soft rot and they are found only in some of the basidiomycetes causing the brown rot, such as *Fomitopsis palustris*[141].

Crystalline cellulose (Avicel, bacterial cellulose or filter paper), which is the main form of cellulose in most plant cell walls are good substrates for exoglucanase activity assay, because it has a low DP and relatively low accessibility; however, some endoglucanases can release considerable reducing sugars from Avicel [13].

# β-GLUCOSIDASES

β-D-glucosidases hydrolyse soluble cellobiose and other cellodextrins with a DP up to 6 to produce glucose in the aqueous phase in order to eliminate cellobiose inhibition [13].

These enzymes have molecular weights ranging from 35 to 640 kDa, and they can be monomeric or exist as homo-oligomers, as is the case β-glucosidase of the yeast *Rhodotorula minuta* [142]. Most β-glucosidases are glycosylated; in some cases, as that of the 300 kDa BGL from *Trametes versicolor*, glycosylation may be superior to 90%. Their optimum pH ranges from 3.5 to 5.5, and their optimum temperature ranges from 45 to 75 °C (3). β-D-glucosidase activities can be measured using cellobiose, which is not hydrolysed by endoglucanases and exoglucanases [13].

# SYNERGY BETWEEN CELLULASES

Synergistic cooperation between cellulases is a prerequisite for efficient degradation of cellulose, but its molecular mechanisms are not fully understood. Synergistic action has been observed between two different cellobiohydrolases and between endoglucanases. However, more synergistic mechanisms have been proposed [143-144]:

Synergy endo-exo, occurs between endo and exoglucanases, where the action of endoglucanases provide free ends of the cellulose chain to the exoglucanases.

Synergy exo-exo, exoglucanases progressively act on reducing and non-reducing ends of the cellulose chain.

Synergy between exoglucanases and β-glucosidases, the latter process cellobiose produced as final product of the action of the exoglucanases.

Intramolecular synergy between catalytic domain and cellulose binding domain of cellulases.

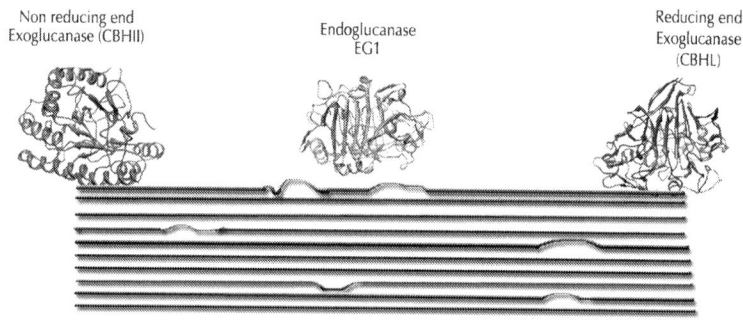

**Figure 7:** Cellulases activities. Exoglucanases act on reducing and non-reducing ends degrading crystalline cellulose, while Endoglucanase act on amorphous cellulose. Structures: CBHI (PBD, 1CB2), CBHII (PDB, 3CBH) and EGL (PDB, 1EG1).

As a whole system, plant cell wall polysaccharides should be degraded efficiently not only by synergy between cellulases but with participation of the other degrading enzymes as xylanases.

In (145) a synergistic mechanism between cellulases and xylanases in order to saccharify wheat straw for bioethanol production is reported. More recently, a new type of synergism between enzymes that employ oxidative reactions to break glycosidic bonds and hydrolytic enzymes was reported in chitin degradation [28].

Although a significant amount of information has been generated related to the action of cellulases and their mechanisms to degrading cellulose, the biodegradation of crystalline cellulose is still a slow process because the substrate is insoluble and poorly accessible to enzymes.

To overcome this situation scientists have optimized ratio of cellulolytic enzymes, and it was found that the best saccharification of crystalline cellulose is achieved with the enzyme blend: 60:20:20 (CBHI:CBHII:EGI) wherein a saturated level of BG was included to eliminate cellobiose inhibition [146]. In a different report, the impact of the cellulase mixture composition on cellulose conversion was modelled, and the findings suggested different optimum ratios for substrates with different characteristics, specifically degrees of polymerization and surface area [147].

Also, researchers have pointed out the use of proteins that relax plant cell wall structure as a complementary activity before action of cellulases in order to improve saccharification.

# PLANT CELL REMODELLING PROTEINS

In addition to lignocellulose-degrading enzymes, there are also enzymes involved in remodelling the cell wall, which could facilitate its later degradation.

## Expansins

Expansins are pH-dependent wall-loosening proteins required for cell enlargement and expansion in many developmental processes. Although to date their precise mechanism of action remains unclear, evidence point toward a role in dissociating the cell wall polysaccharide complex that links together wall components, thus promoting slippage between wall polymers and, eventually, expansion in cell wall [148-149].

These proteins are coded by large multigene families present from bryophytes to angiosperms and also present in monocotyledonous plants (rice, maize), dicotyledonous plants (*Arabidopsis*), ferns and mosses.

Expansins have no hydrolytic activity (glucosidase) and therefore, it has been suggested to work by breaking hydrogen bonds between cellulose fibres or between cellulose and other polysaccharides (xyloglucans), using a non-enzymatic mechanism (Figure 8) [150-153].

Expansins have molecular weights ranging from 25 to 28 kDa and, like cellulases, have a two-domain modular structure and an approximately 20 amino acids-long amino-terminal signal peptide [149].

*Domain I* occupies the N-terminal part of the protein, and it has a DPBB (Double Psi Beta Barrel) structure. It is homologous to the catalytic domain of members of glycoside hydrolase family 45 (GH45), which includes mainly β-1,4-endoglucanases of fungal origin. The

DPBB domain of members of this family adopts a six-stranded beta barrel structure forming a substrate-binding groove. Despite the presence of the GH45 catalytic domain in expansins, no hydrolytic activity has been detected for the latter [5].

*Domain II*, at the C-terminal end, is homologous to group II pollen allergens from grasses. Some authors have speculated that this might be a polysaccharide-binding domain, due to the presence of aromatic and polar amino acids on the protein surface, where two tryptophan and one tyrosine would form a planar platform of aromatic residues favouring this binding (149, 154). Domain II folds as a β-sandwich formed by two sheets of four antiparallel β strands each (Figure 8). In fact, a β-sandwich formed by 3 to 6 β strands per sheet is the most common fold in carbohydrate-binding modules of proteins binding substrates such as crystalline cellulose or chitin [155].

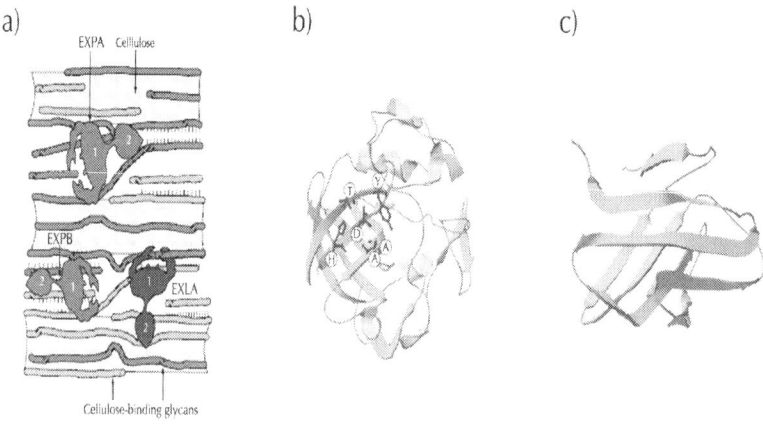

**Figure 8:** a) Expansin proposed activity; b) Expansin domain I (the catalytic domain of a GH45 endoglucanase from *Humicola insolens*; PBD, 2ENG); c) Expansin domain II

(a G2A protein from *Phleum pratense*; PDB 1WHO). In (a), the domain I forms a barrel; amino-acid residues that are conserved in expansins are indicated in the single-letter amino-acid code. Images reproduced with publisher BioMed permission [5].

Expansins are classified in four families: α-expansins (EXPA), β-expansins (EXPB), α-expansin like-proteins (EXLA) and β-expansin like-proteins (EXLB) [5].

The *EXPA family* includes proteins participating in the relaxation and extension of plant cell walls through a pH-dependent mechanism; these proteins would participate in developmental processes such as organogenesis, the degradation of cell walls during the ripening of fruits and other processes where relaxation of the cell wall is crucial [156-159].

The *EXPB family* includes group I pollen allergens from grasses. These proteins are secreted by pollen and have been suggested to soften the tissues of the stigma and style to facilitate the penetration of the pollen tube [154].

EXPB proteins, unlike EXPA members, relax specifically the cell walls of grass cells, probably reflecting differences regarding the organization of cell walls between grasses and dicotyledonous plants. Although an HFD motif, that is known to form part of the active site of endoglucanases, has been found in domain I of EXPA and EXPB family members, they do not have hydrolytic activity [5,160].

The *EXLA and EXLB families* do not have this sequence motif, which suggests that their mode of action differs to that of the other expansins. The EXLA and EXLB families are comprised of proteins identified by sequence analysis which, despite possessing the two- organization typical of expansins, have a number of divergent sequence features that separate them from the EXPA and EXPB families [161].

Another group included in the expansin superfamily is the *expansin-like X family* (EXLX), comprising proteins that exhibit weak sequence homology with the domains of EXPA and EXPB members, and identified in organisms other than plants, such as the mucilaginous fungus *Dictyostelium* and the bacteria *Bacillus subtilis,* and *Hahella chejuensis* [161-164].

The denomination of expansin or expansin-like is reserved for proteins exhibiting both domain I and domain II. Proteins with only one of these domains are not classified as expansins [161]. However other proteins with similar disrupting activity of the cell wall have been reported.

Expansins and expansin-like proteins have been detected in angiosperms such as *Arabidopsis thaliana, Oryza sativa, Zea mays* and *Triticum aestivum,* gymnosperms such as pine and poplar, ferns such as *Regnellidium diphyllum* and *Marsilea quadrifolia* and the moss *Physcomitrella patens*. Some members of the expansin superfamily have been found even in a potato-infecting nematode, *Globodera rostochiensis*, where they are hypothesized to promote the infection process [165-169].

# Novel Proteins with Expansin-Like Activity

Proteins with expansin-like activity called swollenins and loosenins have been identified in ascomycete and basidiomycete fungi such as *T. reesei, A. fumigatus* and *B. adusta* [6-8, 170].

In [7], a swollenin gene from *T. reesei* denominated *swo1,* was cloned and expressed in *Saccharomyces cerevisiae,* coding for a protein that modifies the structure of cellulose in swollen regions of cotton fibres (hence the name) without releasing reducing sugars. Swo1 is a fungal expansin-like protein, containing a pollen allergen domain and a cellulose-binding domain.

Proteins with expansin activity could be used to improve the efficiency of cellulose bioconversion processes. For example, a swollenin purified from *A. fumigatus* has been used in combination with cellulases to facilitate the saccharification of microcrystalline cellulose (Avicel) [8]. In [163] also is described the synergism of an EXLX from *B. subtilis* in the enzymatic hydrolysis of cellulose and recently, and a new protein with expansin activity from the basidiomycete fungus *B. adusta,*denominated loosenin (LOOS1] was cloned and characterized [6].

Not only expansins, but also swollenins and loosenin represent good candidate as pretreatment to enhance sugar production from plant biomass. For example, loosenin activity was efficient to release reducing sugars (after cellulase treatment) from *Agave tequilana,* a crop extensively grown in some areas of Mexico, which shredded fibrous waste is usually burnt or left to decompose. Indeed, *A. tequilana* fiber became a susceptible substrate for a cocktail of commercial cellulases and xylanases in the presence of LOOS1. Loosenin shows optimum activity at the same pH as most cellulolytic enzymes, opening the possibility to use them as a mixture. This protein is able to relax the structure of cotton, enhancing up to 7.5-fold the rate of release of reducing sugars from agave fibre. Something similar was observed when a cucumber expansin was incubated with a compound of cellulose and xyloglucans of bacterial origin and occurred a rapid relaxation of the structure of this compound, suggesting that expansins modulate the binding between cellulose fibres and xyloglucans, relaxing or breaking the bonds keeping them together [171].

Given the optimum pH of LOOS1 (pH 5) and other expansin like proteins, they could be applied to processes of saccharification of natural substrates, facilitating the release of reducing sugars together with cellulases. For example, it might be used as an additive to obtain fermentable sugars from pretreated yellow poplar as reported in [172].

In [173], used swollenin as a pretreatment of cellulosic substrates and observed that even in non-saturating concentrations, a significant accelerated hydrolysis occurred. They also correlated particle size and crystallinity of the cellulosic substrates with initial hydrolysis rates, and it could be shown that the swollenin induced-reduction in particle size and crystallinity resulted in high cellulose hydrolysis rates.

It is not surprising that the idea of using plant expansins in saccharification processes has been patented [174-176].

The efficient enzymatic saccharification of cellulose has been a challenge over the past 50 years, mainly due to its crystallinity, which make it a recalcitrance substrate with a high potential to be used as a carbon source.

The bioconversion of cellulose to ethanol is the process where most interest has been focused. Fortunately, increasing of the loosened cellulose surface area by the use of non-hydrolytic proteins, a process known amorphogenesis, would allow access to hydrolytic enzymes making the saccharification process more efficient [177].

# CONCLUSIONS

Cellulose biodegradation represents the major carbon flow from fixed carbon sinks to atmospheric $CO_2$, this process is very important in several agricultural and waste treatment processes. Also, cellulose contained in plant wastes could be used as a raw material to produce sustainable products and bioenergy to replace depleting fossil fuels. However, one of the most important and difficult technological challenges is to overcome the recalcitrance of natural cellulosic materials, which must be enzymatically hydrolysed to produce fermentable sugars. In order to achieve this goal, new enzymes with cellulolytic activities are being improved and organisms with novel properties have been found. Although the efforts are being directed to improve cellulolytic activity, proteins capable to relax plant cell structure (expansins, swollenins and loosenin) could be used as a biological pretreatment since they would be disrupting crystalline structure of cellulose making it more accessible to the enzymes and enhancing sugar releasing.

# REFERENCES

1.    Lynd LR, Weimer PJ, Van Zyl WH, Pretorius IS. Microbial cellulose utilization: fundamentals and biotechnology. Microbiology and molecular biology reviews. 2002;66(3):506-77.

2.    Baldrian P, Valášková V. Degradation of cellulose by basidiomycetous fungi. FEMS Microbiology Reviews. 2008;32(3):501-21.

3.    Dashtban M, Schraft H, Qin W. Fungal Bioconversion of Lignocellulosic Residues; Opportunities & Perspectives. International Journal of Biological Sciences. 2009;5(6):578-94.

4.    Wada M, Nishiyama Y, Chanzy H, Forsyth T, Langan P. The structure of celluloses. Powder Diffr. 2008;23, No. 2, (2):92-5.

5.    Sampedro J, Cosgrove DJ. The expansin superfamily. Genome biology. 2005;6(12):242.

6.    Quiroz-Castañeda R, Martinez-Anaya C, Cuervo-Soto L, Segovia L, Folch-Mallol J. Loosenin, a novel protein with cellulose-disrupting activity from Bjerkandera adusta. Microbial Cell Factories. 2011;10(1):8.

7.    Saloheimo M, Paloheimo M, Hakola S, Pere J, Swanson B, Nyyssonen E, Bhatia A, Ward M, Penttila M. Swollenin, a *Trichoderma reesei* protein with sequence similarity to the plant expansins, exhibits disruption activity on cellulosic materials. European Journal of Biochemistry. 2002 Sep;269(17):4202-11.

8.    Chen X-a, Ishida N, Todaka N, Nakamura R, Maruyama J, Takahashi H, Kitamoto K. Promotion of Efficient Saccharification with *Aspergillus fumigatus* AfSwo1 Towards Crystalline Cellulose. Applied and Environmental Microbiology. 2010;76(8):2556-61.

9.    Pauly M, Keegstra K. Cell-wall carbohydrates and their modification as a resource for biofuels. The Plant Journal. 2008;54(4):559-68.

10. Field CB, Behrenfeld MJ, Randerson JT, Falkowski P. Primary Production of the Biosphere: Integrating Terrestrial and Oceanic Components. Science. 1998;281(5374):237-40.

11. Lieth H. Primary production of the major vegetation units of the world. In: Primary Productivity of the Biosphere In: Lieth H, Whittaker R, editors.: Springer-Verlag, New York and Berlin. ; 1975. p. 203-15. .

12. Ragauskas AJ, Williams CK, Davison BH, Britovsek G, Cairney J, Eckert CA, Frederick WJ, Jr., Hallett JP, Leak DJ, Liotta CL, Mielenz JR, Murphy R, Templer R, Tschaplinski T. The Path Forward for Biofuels and Biomaterials. Science. 2006 January 27, 2006;311(5760):484-9.

13. Zhang YHP, Lynd LR. Toward an aggregated understanding of enzymatic hydrolysis of cellulose: Noncomplexed cellulase systems. Biotechnology and Bioengineering. 2004;88(7):797-824.

14. Holtzapple MT, in J., eds. Cellulose, Encyclopedia of Food Science, Food Technology, and Nutrition. Macrae R, Robinson, R. K., and Sadler, M. , editor: Academic Press, London, San Diego, CA, NY, Boston, MA, Sydney, Tokio, Toronto, ; 1993.

15. Reddy N, Yang Y. Preparation and characterization of long natural cellulose fibers from wheat straw. J Agric Food Chem. 2007;55(21):8570-5. .

16. Reddy N, Yang Y. Natural cellulose fibers from soybean straw. Bioresour Technol. 2009;100(14):3593-8. .

17. Sun JX, Sun XF, Sun RC, Su YQ. Fractional extraction and structural characterization of sugarcane bagasse hemicelluloses. Carbohydrate Polymers. 2004;56(2):195-204.

18. Fry S. Plant cell walls Encyclopedia of life sciences. 2001;DOI 10.1038/npg.els.0001682. Chichester: Nature Publishing Group.

19. Levy I, Shani Z, Shoseyov O. Modification of polysaccharides and plant cell wall by endo-1,4-[beta]-glucanase and cellulose-binding domains. Biomolecular Engineering. 2002;19(1):17-30.

20. Fernandes AN, Thomas LH, Altaner CM, Callow P, Forsyth VT, Apperley DC, Kennedy CJ, Jarvis MC. Nanostructure of cellulose microfibrils in spruce wood. Proceedings of the National Academy of Sciences USA. 2011;108(47):1195-203.

21.   Endler A, Persson S. Cellulose Synthases and Synthesis in Arabidopsis. Molecular Plant. 2011;4(2):199-211.

22.   Festucci-Buselli RA, Otoni WC, Joshi CP. Structure, organization, and functions of cellulose synthase complexes in higher plants. Brazilian Journal of Plant Physiology. 2007;19:1-13.

23.   Koyama M, Helbert W, Imai T, Sugiyama J, Henrissat B. Parallel-up structure evidences the molecular directionality during biosynthesis of bacterial cellulose. Proceedings of the National Academy of Sciences USA. 1997;94(17):9091-5.

24.   Gautam SP, Bundela PS, Pandey AK, Jamaluddin, Awasthi MK, Sarsaiya S. Diversity of Cellulolytic Microbes and the Biodegradation of Municipal Solid Waste by a Potential Strain. International Journal of Microbiology. 2012;2012.

25.   Kimura S. Immunogold labeling of rosette terminal cellulose-synthesizing complexes in the vascular plant *Vigna angularis*. Plant Cell. 1999;11:2075-85.

26.   Park S, Baker JO, Himmel ME, Parilla PA, Johnson DK. Cellulose crystallinity index: measurement techniques and their impact on interpreting cellulase performance. Biotechnol Biofuels. 2010;3:10.

27.   Horn SJ, Vaaje-Kolstad G, Westereng B, Eijsink VG. Novel enzymes for the degradation of cellulose. Biotechnol Biofuels. 2012;5(1):45.

28.   Vaaje-Kolstad G, Westereng B, Horn SJ, Liu Z, Zhai H, Sørlie M, Eijsink VGH. An Oxidative Enzyme Boosting the Enzymatic Conversion of Recalcitrant Polysaccharides. Science. 2010;330(6001):219-22.

29.   Forsberg Z, Vaaje-Kolstad G, Westereng B, Bunaes AC, Stenstrom Y, MacKenzie A, Sorlie M, Horn SJ, Eijsink VG. Cleavage of cellulose by a CBM33 protein. Protein Sci. 2011;20(9):1479-83. .

30.   Fry SC. Cell Wall Polysaccharide Composition and Covalent Crosslinking. Annual Plant Reviews: Wiley-Blackwell; 2010. p. 1-42.

31.   Hildén L, Johansson G. Recent developments on cellulases and carbohydrate-binding modules with cellulose affinity. Biotechnology Letters. 2004;26(22):1683-93.

32. Percival Zhang YH, Himmel ME, Mielenz JR. Outlook for cellulase improvement: Screening and selection strategies. Biotechnology Advances. 2006;24(5):452-81.

33. Jager G, Wu Z, Garschhammer K, Engel P, Klement T, Rinaldi R, Spiess A, Buchs J. Practical screening of purified cellobiohydrolases and endoglucanases with alpha-cellulose and specification of hydrodynamics. Biotechnology for Biofuels. 2010;3(1):18.

34. Wilkie JS. Carl Nageli and the fine Structure of Living Matter. Nature. 1961;190(4782):1145-50.

35. Meyer KH, Misch L. Positions des atomes dans le nouveau modèle spatial de la cellulose. Helvetica Chimica Acta. 1937;20(1):232-44.

36. Atalla RH, Vanderhart DL. Native Cellulose: A Composite of Two Distinct Crystalline Forms. Science. 1984;223(4633):283-5.

37. Kovalenko VI. Crystalline cellulose: structure and hydrogen bonds. Russian Chemical Reviews. 2010;79(3):231-41.

38. Li Y, Lin M, Davenport JW. Ab initio studies of cellulose I: crystal structure, intermolecular forces, and interactions with water. The journal of physical chemistry. 2011;115:11533-9.

39. Klemm D, Schmauder HP, Heinze T. Cellulose. In: Steinbüchel A, editor. Biopolymers Volume 6Polysaccharides II: Polysaccharides from Eukaryotes Münster, Germany: Wiley-VCH; 2004.

40. Zugenmaier P. Conformation and packing of various crystalline cellulose fibers. Progress in Polymer Science. 2001;26(9):1341-417.

41. Nishiyama Y, Sugiyama J, Chanzy H, Langan P. Crystal structure and hydrogen bonding system in cellulose I(alpha) from synchrotron X-ray and neutron fiber diffraction. J Am Chem Soc. 2003;125(47):14300-6.

42. Nishiyama Y, Langan P, Chanzy H. Crystal structure and hydrogen-bonding system in cellulose Ibeta from synchrotron X-ray and neutron fiber diffraction. J Am Chem Soc. 2002;124(31):9074-82.

43. Langan P, Nishiyama Y, Chanzy H. X-ray Structure of Mercerized Cellulose II at 1 Å Resolution. Biomacromolecules. 2001;2(2):410-6.

44. Langan P, Sukumar N, Nishiyama Y, Chanzy H. Synchrotron X-ray structures of cellulose Iβ; and regenerated cellulose II at ambient temperature and 100 K. Cellulose. 2005;12(6):551-62.

45. Wada M, Chanzy H, Nishiyama Y, Langan P. Cellulose IIII Crystal Structure and Hydrogen Bonding by Synchrotron X-ray and Neutron Fiber Diffraction. Macromolecules. [doi: 10.1021/ma0485585]. 2004;37(23):8548-55.

46. Wada M, Heux L, Sugiya J. Polymorphism of cellulose I family: Reinvestigation of cellulose IVI. Biomacromolecules 2004;5:1385-91.

47. Mittal A, Katahira R, Himmel M, Johnson D. Effects of alkaline or liquid-ammonia treatment on crystalline cellulose: changes in crystalline structure and effects on enzymatic digestibility. Biotechnology for Biofuels. 2011;4(1):41.

48. Hall M, Bansal P, Lee JH, Realff MJ, Bommarius AS. Cellulose crystallinity – a key predictor of the enzymatic hydrolysis rate. FEBS Journal. 2010;277(6):1571-82.

49. Gardiner ES, Sarko A. Packing analysis of carbohydrates and polysaccharides. 16. The crystal structures of cellulose IVI and IVII. CanJ Chemistry. 1985;63:173-80. .

50. Somerville C, Bauer S, Brininstool G, Facette M, Hamann T, Milne J, Osborne E, Paredez A, Persson S, Raab T, Vorwerk S, Youngs H. Toward a Systems Approach to Understanding Plant Cell Walls. Science. 2004 December 24, 2004;306(5705):2206-11.

51. Ding S-Y, Himmel ME. The Maize Primary Cell Wall Microfibril: A New Model Derived from Direct Visualization. Journal of Agricultural and Food Chemistry. 2006;54(3):597-606.

52. Doi R. Cellulases of mesophilic microorganisms: cellulosome & non-cellulosome producers. Annals of the New York Academy of Sciences. 2008;1125:267-79.

53. Ransom-Jones E, Jones D, McCarthy A, McDonald J. The Fibrobacteres: an important phylum of cellulose-degrading bacteria. Microb Ecol 2012 Feb;63(2):267-81 2012.

54. Schwarz W. The cellulosome and cellulose degradation by anaerobic bacteria. Applied Microbiology and Biotechnology. 2001;56:634–49.

55. Dillon RJ, Dillon VM. The gut bacteria of insects: nonpathogenic interactions. Annu Rev Entomol. 2004;49:71-92.

56. Arcand N, Kluepfel D, Paradis F, Morosoli R, Shareck F. Beta-mannanase of *Streptomyces lividans* 66: cloning and DNA sequence of the manA gene and characterization of the enzyme. Biochemical Journal. 1993;290(3):857-63.

57. Khanna S, Gauri. Regulation, purification, and properties of xylanase from *Cellulomonas fimi*. Enzyme and Microbial Technology. 1993;15(11):990-5.

58. Braithwaite KL, Black GW, Hazlewood GP, Ali BR, Gilbert HJ. A non-modular endo-beta-1,4-mannanase from *Pseudomonas fluorescens* subspecies cellulosa. Biochemical Journal. 1995;305(3):1005-10.

59. Lin L, Thomson J. An analysis of the extracellular xylanases and cellulases of *Butyrivibrio fibrisolvens* H17c. FEMS Microbiology Letters. 1991;84(2):197-204.

60. Murty MVS, Chandra TS. Purification and properties of an extra cellular xylanase enzyme of *Clostridium* strain SAIV. Antonie van Leeuwenhoek. 1992;61(1):35-41.

61. Tomme P, Warren R, Gilke N. Cellulose hydrolysis by bacteria and fungi. Advances In Microbial Physiology. 1995;37:1–81.

62. Eberhardt RY, Gilbert HJ, Hazlewood GP. Primary sequence and enzymic properties of two modular endoglucanases, Cel5A and Cel45A, from the anaerobic fungus *Piromyces equi*. Microbiology. 2000;146(8):1999-2008.

63. Sánchez C. Lignocellulosic residues: Biodegradation and bioconversion by fungi. Biotechnology Advances. 2009;27(2):185-94.

64. Wilson DB. Microbial diversity of cellulose hydrolysis. Curr Opin Microbiol. 2011;14(3):259-63. Epub 2011 Apr 29.

65. Quiroz-Castañeda RE, Balcazar-Lopez E, Dantan-Gonzalez E, Martinez A, Folch-Mallol J, Martinez-Anaya C. Characterization of cellulolytic activities of *Bjerkandera adusta* and *Pycnoporus sanguineus* on solid wheat straw medium. Electronic Journal of Biotechnology [online]. 2009;12(4):Available from Internet: http://www.ejbiotechnology.cl/content/vol12/issue4/full/3/index.html.

66. Koseki T, Yuichiro M, Shinya F, Kazuo M, Tsutomu F, Kiyoshi I, Yoshihit S, Haruyuki I. Biochemical characterization of a glycoside hydrolase family 61 endoglucanase from *Aspergillus kawachii*. Applied Microbiology and Biotechnology. 2008;77:1279–85.

67. Lamed R, Naimark J, Morgenstern E, Bayer EA. Specialized cell surface structures in cellulolytic bacteria. J Bacteriol. 1987;169(8):3792-800.

68. Kurzatkowski W, Torronen A, Filipek J, Mach RL, Herzog P, Sowka S, Kubicek CP. Glucose-induced secretion of *Trichoderma reesei* xylanases. Appl Environ Microbiol 1996;62(8):2859-65.

69. Chao Y, Singh D, Yu L, Li Z, Chi Z, Chen S. Secretome characteristics of pelletized Trichoderma reesei and cellulase production. World J Microbiol Biotechnol. 2012;28(8):2635-41. Epub 012 May 12.

70. Do Vale LH, Gomez-Mendoza DP, Kim MS, Pandey A, Ricart CA, Ximenes-Filho E, Sousa MV. Secretome analysis of the fungus Trichoderma harzianum grown on cellulose. Proteomics. 2012;29(10):201200063.

71. Garcia-Kirchner O, Segura-Granados M, Rodriguez-Pascual P. Effect of media composition and growth conditions on production of beta-glucosidase by Aspergillus niger C-6. Appl Biochem Biotechnol. 2005;124:347-59.

72. Tsujiyama S, Ueno H. Production of cellulolytic enzymes containing cinnamic acid esterase from Schizophyllum commune. J Gen Appl Microbiol. 2011;57(6):309-17.

73. Ray A, Saykhedkar S, Ayoubi-Canaan P, Hartson SD, Prade R, Mort AJ. Phanerochaete chrysosporium produces a diverse array of extracellular enzymes when grown on sorghum. Appl Microbiol Biotechnol. 2012;93(5):2075-89.

74. Quiroz-Castañeda R, Pérez-Mejía N, Martínez-Anaya C, Acosta-Urdapilleta L, Folch-Mallol J. Evaluation of different lignocellulosic substrates for the production of cellulases and xylanases by the basidiomycete fungi *Bjerkandera adusta* and *Pycnoporus sanguineus*. Biodegradation. 2010:1-8.

75. Ji HW, Cha CJ. Identification and functional analysis of a gene encoding beta-glucosidase from the brown-rot basidiomycete Fomitopsis palustris. J Microbiol. 2010;48(6):808-13. Epub 2011 Jan 9.

76.  Lee SS, Ha JK, Cheng KJ. The effects of sequential inoculation of mixed rumen protozoa on the degradation of orchard grass cell walls by anaerobic fungus Anaeromyces mucronatus 543. Can J Microbiol. 2001;47(8):754-60.

77.  Hodrova B, Kopecny J, Kas J. Cellulolytic enzymes of rumen anaerobic fungi Orpinomyces joyonii and Caecomyces communis. Res Microbiol. 1998;149(6):417-27.

78.  Griffith GW, Ozkose E, Theodorou MK, Davies DR. Diversity of anaerobic fungal populations in cattle revealed by selective enrichment culture using different carbon sources. Fungal Ecology. [doi: 10.1016/j.funeco.2009.01.005]. 2009;2(2):87-97.

79.  Li XL, Chen H, Ljungdahl LG. Monocentric and polycentric anaerobic fungi produce structurally related cellulases and xylanases. Appl Environ Microbiol. 1997;63(2):628-35.

80.  Chen H-L, Chen Y-C, Lu M-Y, Chang J-J, Wang H-T, Wang T-Y, Ruan S-K, Wang T-Y, Hung K-Y, Cho H-Y, Ke H-M, Lin W-T, Shih M-C, Li W-H. A highly efficient beta-glucosidase from a buffalo rumen fungus Neocallimastix patriciarum W5. Biotechnology for Biofuels. 2012;5(1):24.

81.  Anderson I, Abt B, Lykidis A, Klenk H-P, Kyrpides N, Ivanova N. Genomics of Aerobic Cellulose Utilization Systems in Actinobacteria. PLoS ONE. 2012;7(6).

82.  Freier D MC, Wiegel J: . Characterization of *Clostridium thermocellum* JW20. . Appl Environ Microbiol 1988;54:204-11.

83.  Tuncer M, Ball AS. Degradation of lignocellulose by extracellular enzymes produced by Thermomonospora fusca BD25. Appl Microbiol Biotechnol. 2002;58(5):608-11. .

84.  Bredholt S, Sonne-Hansen J, Nielsen P, Mathrani IM, Ahring BK. Caldicellulosiruptor kristjanssonii sp. nov., a cellulolytic, extremely thermophilic, anaerobic bacterium. Int J Syst Bacteriol. 1999;3:991-6.

85.  Svetlichnyi V, Svetlichnaya T, Chernykh N, Zavarzin G. *Anaerocellum thermophilum* gen. nov., sp. nov., an extremely thermophilic cellulolytic eubacterium isolated from hot-springs in the valley of Geysers. . Microbiology. 1990;59:598–604.

86.  Yang SJ, Kataeva I, Hamilton-Brehm SD, Engle NL, Tschaplinski TJ, Doeppke C, Davis M, Westpheling J, Adams MW. Efficient

degradation of lignocellulosic plant biomass, without pretreatment, by the thermophilic anaerobe "Anaerocellum thermophilum" DSM 6725. Appl Environ Microbiol. 2009;75(14):4762-9. .

87. Doi R, Kosugi A. Cellulosomes: plant cell wall degrading enzyme complexes. Nature reviews microbiology. 2004;2:541-51.

88. Bayer EA, Belaich JP, Shoham Y, Lamed R. The cellulosomes: multienzyme machines for degradation of plant cell wall polysaccharides. Annu Rev Microbiol. 2004;58:521-54.

89. Blumer-Schuette SE, Kataeva I, Westpheling J, Adams MW, Kelly RM. Extremely thermophilic microorganisms for biomass conversion: status and prospects. Curr Opin Biotechnol. 2008;19(3):210-7.

90. Fontes CMGA, Gilbert HJ. Cellulosomes: Highly Efficient Nanomachines Designed to Deconstruct Plant Cell Wall Complex Carbohydrates. Annual Review of Biochemistry. 2010;79(1):655-81.

91. Sun Y, Cheng J. Hydrolysis of lignocellulosic materials for ethanol production: a review. Bioresource Technology. 2002;83(1):1-11.

92. Henrissat B, Teeri TT, Warren RA. A scheme for designating enzymes that hydrolyse the polysaccharides in the cell walls of plants. FEBS Lett. 1998;425(2):352-4.

93. Kuhad RC, Gupta R, Singh A. Microbial Cellulases and Their Industrial Applications. Enzyme Research. 2011;2011.

94. Din N, Gilkes NR, Tekant B, Miller RC, Warren RAJ, Kilburn DG. Non-Hydrolytic Disruption of Cellulose Fibres by the Binding Domain of a Bacterial Cellulase. Nat Biotech. [10.1038/nbt1191-1096]. 1991;9(11):1096-9.

95. Irwin D, Spezio M, Walker L, DB W. Activity studies of eight purified cellulases: specificity, synergism, and binding domain effects. Biotechnology and Bioengineering. 1993;42:1002–13.

96. Feller G, Gerday C. Psychrophilic enzymes: hot topics in cold adaptation. Nat Rev Microbiol. 2003;1(3):200-8.

97. Sonan G, Receveur-Brechot V, Duez C, Aghajari N, Czjzek M, Haser R, Gerday C. The linker region plays a key role in the adaptation to cold of the cellulase from an Antarctic bacterium. Biochemical Journal. 2007;407:293–302.

98.  Maki M, Leung KT, Qin W. The prospects of cellulase-producing bacteria for the bioconversion of lignocellulosic biomass. Int J Biol Sci. 2009;5(5):500-16.

99.  Li DC, Li AN, Papageorgiou AC. Cellulases from thermophilic fungi: recent insights and biotechnological potential. Enzyme Res. 2011;2011:308730.

100. Miroshnichenko ML, Kublanov IV, Kostrikina NA, Tourova TP, Kolganova TV, Birkeland NK, Bonch-Osmolovskaya EA. Caldicellulosiruptor kronotskyensis sp. nov. and Caldicellulosiruptor hydrothermalis sp. nov., two extremely thermophilic, cellulolytic, anaerobic bacteria from Kamchatka thermal springs. Int J Syst Evol Microbiol. 2008;58(Pt 6):1492-6.

101. Grogan DW. Evidence that beta-Galactosidase of Sulfolobus solfataricus Is Only One of Several Activities of a Thermostable beta-d-Glycosidase. Appl Environ Microbiol. 1991;57(6):1644-9.

102. Zambare VP, Bhalla A, Muthukumarappan K, Sani RK, Christopher LP. Bioprocessing of agricultural residues to ethanol utilizing a cellulolytic extremophile. Extremophiles. 2011;15(5):611-8. .

103. Kang HJ, Uegaki K, Fukada H, Ishikawa K. Improvement of the enzymatic activity of the hyperthermophilic cellulase from Pyrococcus horikoshii. Extremophiles. 2007;11(2):251-6.

104. Bauer MW, Driskill LE, Callen W, Snead MA, Mathur EJ, Kelly RM. An endoglucanase, EglA, from the hyperthermophilic archaeon Pyrococcus furiosus hydrolyzes beta-1,4 bonds in mixed-linkage (1-->3),(1-->4)-beta-D-glucans and cellulose. J Bacteriol. 1999;181(1):284-90.

105. Dutta T, Sahoo R, Sengupta R, Ray SS, Bhattacharjee A, Ghosh S. Novel cellulases from an extremophilic filamentous fungi Penicillium citrinum: production and characterization. J Ind Microbiol Biotechnol. 2008;35(4):275-82. .

106. Kaper T, Brouns SJJ, Geerling ACM, De Vos WM, Van der Oost J. DNA family shuffling of hyperthermostable beta-glycosidases. Biochem J. 2002;368(2):461-70.

107. Rees HC, Grant S, Jones B, Grant WD, Heaphy S. Detecting cellulase and esterase enzyme activities encoded by novel genes present in environmental DNA libraries. Extremophiles. 2003;7(5):415-21. .

108. Ito S. Alkaline cellulases from alkaliphilic Bacillus: enzymatic properties, genetics, and application to detergents. Extremophiles. 1997;1(2):61-6.

109. Shoseyov O, Shani Z, Levy I. Carbohydrate Binding Modules: Biochemical Properties and Novel Application. Microbiology and molecular biology reviews. 2006:283–95.

110. Bhat M, Bhat S. Cellulose degrading enzymes and their potential industrial applications. Biotechnology Advances. 1997;15:583-620.

111. Davies G, Henrissat B. Structures and mechanisms of glycosyl hydrolases. Structure. 1995;3(9):853-9.

112. Teeri T, Koivula A, Linder M, Wohlfahrt G, Divne C, Jones T. Trichoderma reesei cellobiohydrolases: why so efficient on crystalline cellulose? Biochemical Society Transactions. 1998;26:173–8.

113. Margolles-Clark E, Tenkanen M, Soderlund H, Pentilla.M. Acetyl xylan esterase from Trichoderma reesei contains an active site serine and a cellulose-binding domain. European Journal of Biochemistry. 1996;237:553–60.

114. Wilson D, Irwin D. Genetics and Properties of Cellulases. Advances in Biochemical Engineering / Biotechnology. 1999;65:1-21.

115. Srisodsuk M, Reinikainen T, Penttila M, Teeri T. Role of the interdomain linker peptide of Trichoderma reesei cellobiohydrolase I in its interaction with crystalline cellulose. Journal of Biological Chemistry. 1993;268:20756–61.

116. Shen H, Schmuck M, Pilz I, Gilkes N, Kilburn D, Miller R, Warren R. Deletion of the linker connecting the catalytic and cellulose-binding domains of endoglucanase A (CenA) of Cellulomonas fimi alters its conformation and catalytic activity. Journal of Biological Chemistry. 1991;266:11335–40.

117. Chundawat SP, Beckham GT, Himmel ME, Dale BE. Deconstruction of lignocellulosic biomass to fuels and chemicals. Annu Rev Chem Biomol Eng. 2011;2:121-45.

118. Liu QP, Sulzenbacher G, Yuan H, Bennett EP, Pietz G, Saunders K, Spence J, Nudelman E, Levery SB, White T, Neveu JM, Lane WS, Bourne Y, Olsson ML, Henrissat B, Clausen H. Bacterial

glycosidases for the production of universal red blood cells. Nat Biotechnol. 2007;25(4):454-64. .

119. Rajan SS, Yang X, Collart F, Yip VL, Withers SG, Varrot A, Thompson J, Davies GJ, Anderson WF. Novel catalytic mechanism of glycoside hydrolysis based on the structure of an NAD+/Mn2+-dependent phospho-alpha-glucosidase from Bacillus subtilis. Structure. 2004;12(9):1619-29.

120. Dworkin M, Rosenberg E, Schleifer K. The Prokaryotes: Ecophysiology and biochemistry. New York, USA.: Springer; 2006.

121. Nguyen NH, Maruset L, Uengwetwanit T, Mhuantong W, Harnpicharnchai P, Champreda V, Tanapongpipat S, Jirajaroenrat K, Rakshit SK, Eurwilaichitr L, Pongpattanakitshote S. Identification and characterization of a cellulase-encoding gene from the buffalo rumen metagenomic library. Biosci Biotechnol Biochem. 2012;76(6):1075-84.

122. van der Lelie D, Taghavi S, McCorkle SM, Li L-L, Malfatti SA, Monteleone D, Donohoe BS, Ding S-Y, Adney WS, Himmel ME, Tringe SG. The Metagenome of an Anaerobic Microbial Community Decomposing Poplar Wood Chips. PLoS ONE. [doi:10.1371/journal.pone.0036740]. 2012;7(5):36740.

123. Nimchua T, Thongaram T, Uengwetwanit T, Pongpattanakitshote S, Eurwilaichitr L. Metagenomic analysis of novel lignocellulose-degrading enzymes from higher termite guts inhabiting microbes. J Microbiol Biotechnol. 2012;22(4):462-9.

124. Li LL, McCorkle SR, Monchy S, Taghavi S, van der Lelie D. Bioprospecting metagenomes: glycosyl hydrolases for converting biomass. Biotechnol Biofuels. 2009;2:10.

125. Kellner H, Vandenbol M. Fungi Unearthed: Transcripts Encoding Lignocellulolytic and Chitinolytic Enzymes in Forest Soil. PLoS ONE. 2010;5(6):10971.

126. Li LL, Taghavi S, McCorkle SM, Zhang YB, Blewitt MG, Brunecky R, Adney WS, Himmel ME, Brumm P, Drinkwater C, Mead DA, Tringe SG, Lelie D. Bioprospecting metagenomics of decaying wood: mining for new glycoside hydrolases. Biotechnol Biofuels. 2011;4(1):23.

127. Findley SD, Mormile MR, Sommer-Hurley A, Zhang XC, Tipton P, Arnett K, Porter JH, Kerley M, Stacey G. Activity-based metagenomic screening and biochemical characterization of bovine ruminal protozoan glycoside hydrolases. Appl Environ Microbiol. 2011;77(22):8106-13.

128. Chi Z, Chi Z, Zhang T, Liu G, Li J, Wang X. Production, characterization and gene cloning of the extracellular enzymes from the marine-derived yeasts and their potential applications. Biotechnology Advances. 2009;27:236–55

129. Kikuchi T, Jones J, Aikawa T, Kosaka H, Ogura N. A family of glycosyl hydrolase family 45 cellulases from the pine wood nematode*Bursaphelenchus xylophilus*. FEBS Letters. 2004;572:201–5.

130. Ekborg NA, Gonzalez JM, Howard MB, Taylor LE, Hutcheson SW, Weiner RM. Saccharophagus degradans gen. nov., sp. nov., a versatile marine degrader of complex polysaccharides. Int J Syst Evol Microbiol. 2005;55(Pt 4):1545-9.

131. Beckham GT, Bomble YJ, Bayer EA, Himmel ME, Crowley MF. Applications of computational science for understanding enzymatic deconstruction of cellulose. Current Opinion in Biotechnology. [doi: 10.1016/j.copbio.2010.11.005]. 2011;22(2):231-8.

132. Henrissat B. A classification of glycosyl hydrolases based on amino acid sequence similarities. Biochem J. 1991;280(Pt 2):309-16.

133. Henrissat B, Bairoch A. New families in the classification of glycosyl hydrolases based on amino acid sequence similarities. Biochem J. 1993;293(Pt 3):781-8.

134. Boraston AB, Bolam DN, Gilbert HJ, Davies GJ. Carbohydrate-binding modules: fine-tuning polysaccharide recognition. Biochem J. 2004 382(3):769-81.

135. Aro N, Pakula T, Pentilla M. Transcriptional regulation of plant cell wall degradation by filamentous fungi. FEMS Microbiology Reviews. 2005;29(4):719-39.

136. Lynd L, Cushman J, Nichols R, Wyman C. Fuel ethanol from cellulosic biomass. Science. 1991;15:1318–23.

137. Sadana J, Lachke A, Patil R. Endo-(1-4)-beta-D-glucanases from *Sclerotium rolfsii* – purification, substrate specificity, and mode of action. Carbohydrate Research. 1984;133:297–312.

138. Valásková V, Baldrian P. Degradation of cellulose and hemicelluloses by the brown rot fungus *Piptoporus betulinus* – production of extracellular enzymes and characterization of the major cellulases. Microbiology. 2006;152: 3613–22.

139. Ding S, Ge W, Buswell J. Endoglucanase I from the edible straw mushroom, *Volvariella volvacea*. European Journal of Biochemistry. 2001;268(22):5687-95.

140. Sadana J, Patil R. 1,4-beta-D-glucan cellobiohydrolase from *Sclerotium rolfsii*. Methods in Enzymology. 1988;160: 307–14.

141. Song BC, Kim KY, Yoon JJ, Sim SH, Lee K, Kim YS, Kim YK, Cha CJ. Functional analysis of a gene encoding endoglucanase that belongs to glycosyl hydrolase family 12 from the brown-rot basidiomycete Fomitopsis palustris. J Microbiol Biotechnol. 2008;18(3):404-9.

142. Onishi N, Tanaka T. Purification and properties of a galacto- and gluco-oligosaccharide-producing betaglycosidase from *Rhodotorula minuta*IFO879. . Journal of Fermentation and Bioengineering. 1996;82:439–43.

143. Teeri T. Crystalline cellulose degradation: new insight into the function of cellobiohydrolases Trends in Biotechnology. 1997;15:160–7.

144. Jalak J, Kurashin M, Teugjas H, Valjamae P. Endo-exo synergism in cellulose hydrolysis revisited. J Biol Chem. 2012;25:25.

145. Tabka MG, Herpoël-Gimbert I, Monod F, Asther M, Sigoillot JC. Enzymatic saccharification of wheat straw for bioethanol production by a combined cellulase xylanase and feruloyl esterase treatment. Enzyme and Microbial Technology. 2006;39(4):897-902.

146. Baker JO, Ehrman CI, Adney WS, Thomas SR, Himmel ME. Hydrolysis of cellulose using ternary mixtures of purified celluloses. Appl Biochem Biotechnol. 1998;72:395-403.

147. Levine SE, Fox JM, Clark DS, Blanch HW. A mechanistic model for rational design of optimal cellulase mixtures. Biotechnol Bioeng. 2011;108(11):2561-70. .

148. Cosgrove DJ. Growth of the plant cell wall. Nature Reviews Molecular Cell Biology. 2005;6(11):850-61.

149. Cosgrove DJ. Loosening of plant cell walls by expansins. Nature. 2000;407(6802):321-6.

150. McQueen-Mason S, Cosgrove DJ. Disruption of hydrogen bonding between plant cell wall polymers by proteins that induce wall extension. Proceedings of the National Academy of Sciences USA. 1994;91(14):6574-8.

151. Lee Y, Choi D, Kende H. Expansins: ever-expanding numbers and functions. Current Opinion in Plant Biology. 2001;4(6):527-32.

152. Wei W, Yanga C, Luoa J, Lua C, Wub Y, Yuana S. Synergism between cucumber alpha-expansin, fungal endoglucanase and pectin lyase. Journal of Plant Physiology. 2010;167:1204–10.

153. Li Y, Jones L, McQueen-Mason S. Expansins and cell growth. Current Opinion in Plant Biology. 2003;6(6):603-10.

154. Cosgrove DJ. Relaxation in a high-stress environment: the molecular bases of extensible cell walls and cell enlargement. Plant Cell. 1997;9(7):1031-41.

155. Kerff F, Amoroso A, Herman R, Sauvage E, Petrella S, Filee P, Charlier P, Joris B, Tabuchi A, Nikolaidis N, Cosgrove DJ. Crystal structure and activity of Bacillus subtilis YoaJ (EXLX1), a bacterial expansin that promotes root colonization. Proceedings of the National Academy of Sciences USA. 2008;105(44):16876-81.

156. Rose JKC, Lee HH, Bennett AB. Expression of a divergent expansin gene is fruit-specific and ripening-regulated. Proceedings of the National Academy of Sciences USA. 1997;94(11):5955-60.

157. Civello PM, Powell ALT, Sabehat A, Bennett AB. An Expansin Gene Expressed in Ripening Strawberry Fruit. Plant Physiology. 1999;121(4):1273-9.

158. Cho HT, Cosgrove DJ. Regulation of root hair initiation and expansin gene expression in Arabidopsis. Plant Cell. 2002;14(12):3237-53.

159. Cosgrove D, Li L, Cho H, Hoffmann-Benning S, Moore R, Blecker D. The growing world of expansins. Plant Cell Physiology. 2002 43(12):1436-44.

160. Cosgrove D, Bedinger P, Durachko D. Group I allergens of grass pollen as cell wall-loosening agents. Proceedings of the National Academy of Sciences USA. 1997;94(12):6559-64.

161. Kende H, Bradford K, Brummell D, Cho HT, Cosgrove D, Fleming A, Gehring C, Lee Y, McQueen-Mason S, Rose J, Voesenek LA. Nomenclature for members of the expansin superfamily of genes and proteins. Plant Molecular Biology. 2004;55(3):311-4.

162. Darley CP, Li Y, Schaap P, McQueen-Mason SJ. Expression of a family of expansin-like proteins during the development of *Dictyostelium discoideum*. FEBS Letters. 2003;546(2-3):416-8.

163. Kim ES, Lee HJ, Bang WG, Choi IG, Kim KH. Functional characterization of a bacterial expansin from *Bacillus subtilis* for enhanced enzymatic hydrolysis of cellulose. Biotechnology and Bioengineering. 2009;102(5):1342-53.

164. Lee HJ, Lee S, Ko HJ, Kim KH, Choi IG. An expansin-like protein from Hahella chejuensis binds cellulose and enhances cellulase activity. Molecules and cells. 2010;29(4):379-85.

165. Carey RE, Cosgrove DJ. Portrait of the expansin superfamily in *Physcomitrella patens*: comparisons with angiosperm expansins. Annals of Botany. 2007;99(6):1131-41.

166. Wu Y, Meeley R, Cosgrove D. Analysis and expression of the α-expansin and β-expansin gene families in maize. Plant Physiology. 2001;126:222-32.

167. Li Y, Darley C, Ongaro V, Fleming A, Schipper O, Baldauf S, McQueen-Mason S. Plant expansins are a complex multigene family with an ancient evolutionary origin. Plant Physiology. 2002;128:854–64.

168. Lin Z, Ni Z, Zhang Y, Yao Y, Wu H, Sun Q. Isolation and characterization of 18 genes encoding α- and β-expansins in wheat (*Triticum aestivum*). Molecular Genetics and Genomics. 2005;274(5):548-56.

169. Kudla U, Qin L, Milac A, Kielak A, Maissen C, Overmars H, Popeijus H, Roze E, Petrescu A, Smant G, Bakker J, Helder J. Origin, distribution and 3D-modeling of Gr-EXPB1, an expansin from the potato cyst nematode *Globodera rostochiensis*. FEBS Letters. 2005;579(11):2451-7.

170. Brotman Y, Briff E, Viterbo A, Chet I. Role of Swollenin, an Expansin-Like Protein from *Trichoderma*, in Plant Root Colonization. Plant Physiology. 2008;147(2):779-89.

171. Whitney S, Gidley M, McQueen-Mason S. Probing expansin action using cellulose/hemicellulose composites. Plant Journal. 2000;22:327–34.

172. Baker J, King M, Adney W, Decker S, Vinzant T, Lantz S, Nieves R, Thomas S, Li L-C, Cosgrove D, Himmel M. Investigation of the cell-wall loosening protein expansin as a possible additive in the enzymatic saccharification of lignocellulosic biomass. Applied Biochemistry and Biotechnology. 2000;84-86(1):217-23.

173. Jager G, Girfoglio M, Dollo F, Rinaldi R, Bongard H, Commandeur U, Fischer R, Spiess A, Buchs J. How recombinant swollenin from Kluyveromyces lactis affects cellulosic substrates and accelerates their hydrolysis. Biotechnology for Biofuels. 2011;4(1):33.

174. Cosgrove D, inventor The Penn State Research Foundation., assignee. Enhancement of accessibility of cellulose by expansins. . United States 2001.

175. Cosgrove D, inventor The Penn State Research Foundation, assignee. Increased activity and efficiency of expansin-like proteins. United States2007.

176. Cosgrove D, inventor The Penn State Research Foundation assignee. β-expansins as cell wall loosening agents, compositions thereof and methods of use 2004.

177. Arantes V, Saddler J. Access to cellulose limits the efficiency of enzymatic hydrolysis: the role of amorphogenesis. Biotechnology for Biofuels. 2010;3(1):4.

# Novel Crystalline Sio2 Nanoparticles Via Annelids Bioprocessing of Agro-Industrial Wastes

A Espíndola-Gonzalez[12], AL Martínez-Hernández[3],
C Angeles-Chávez[4], VM Castaño[1], and C Velasco-Santos[3]

[1]Centro de Física Aplicada y Tecnología Avanzada Universidad Nacional Autónoma de México, Campus Juriquilla, Querétaro, México

[2]Facultad de Ingeniería, Universidad Nacional Autónoma de México, Edificio Bernardo Quintana. Cd. Universitaria, Querétaro, 04510, México

[3]Instituto Tecnológico de Querétaro, Av. Tecnológico S/N Esq. Gral. Mariano Escobedo, Col, Centro Histórico, 76000, México

[4]Instituto Mexicano del Petróleo, Eje Central Lázaro Cárdenas Norte 152, San Bartolo Atepehuacan, Gustavo A. Madero, 07730, México

# ABSTRACT

The synthesis of nanoparticles silica oxide from rice husk, sugar cane bagasse and coffee husk, by employing vermicompost with annelids (*Eisenia foetida*) is reported. The product (*humus*) is calcinated and extracted to recover the crystalline nanoparticles. X-ray diffraction (XRD), transmission electron microscopy (TEM), high-resolution transmission electron microscopy (HRTEM) and dynamic light scattering (DLS) show that the biotransformation allows creating specific crystalline phases, since equivalent particles synthesized without biotransformation are bigger and with different crystalline structure.

# INTRODUCTION

Agro-industrial wastes have recently attracted a great deal of attention as potential sources of novel green alternatives such as biotransformation for fuels and other materials. Many of these wastes contain amorphous silica that can be transformed into crystalline nanoparticles of industrial interest.

Silicon is the most common element of the Earth's surface after oxygen; this element is released into the soil by chemical and biological processes [1]. Industrially speaking, silicon is the basis of semiconductors, glasses, ceramics, plastics, elastomers, resins, mesoporous molecular sieves and catalysts, optical fibers and coatings, insulators, moisture shields, photoluminescent polymers, fillers, cosmetics and biomedical devices [2,3], among many other applications. The manufacture of these materials typically requires high temperatures, high pressure and/or the use of caustic chemicals [4]. In contrast, in nature, silica architectures with delicate morphologies are generated under ambient conditions [5]. Unicellular organisms, such as diatoms, use structuring and templating biomolecules to produce silica shells that not only contain hierarchically ordered porous structures, with dimensions ranging from the nanometer to the micrometer domain, but also possess remarkable mechanical and structural properties [6-8]. Other approach involves the use*Fosuarium oxysporum*, a plant pathogenic fungus for the biotransformation of naturally occurring amorphous plant bio-silica into quasi-spherical crystalline silica nanoparticles and its extracellular leaching in the aqueous environment at room temperature [9]. An analysis suggested that extreme thermophilic bacteria within the genera *thermus* and *hydrogenobacter* are predominant components among the indigenous microbial community in siliceous deposits. These bacteria seem to actively contribute to the rapid formation of huge siliceous deposits [10].

In general, the biosilification products are commonly composed of amorphous silica (opal-A, opal-CT and opal-C), and other crystal arrays such as cristobalite, trydimite and quartz. In particular, amorphous silica is a dominant component in marine surface sediment most of which is considered to be generated by the activity of living organism [3,11]. Many scientists not only investigate the process underlying their formation, but also aim to mimic these processes in order to obtain better control over the structure and morphology of chemically produced silica [12-14]. The natural silica production receives increasing attention, since it holds the key to the formation of silica morphologies with a dedicated organization of hierarchically structure elements and the ability to synthesize such silica under ambient conditions [15]. There exist many studies in silica bio-mineralization of simple aquatic life forms, including unicellular organisms like diatoms, radiolaria and sinurophytes as multicellular sponges [16-19]. In the soil, silica plays a major role in higher plants [20]. Many plants sequester silica in biogenic phytoliths and soils can accumulate significant quantities of biogenic opal-A [21]. The silica absorbed for terrestrial plants is around a fraction of 1% of the dry matter to several percent, and in some plants to 10% or even higher [22]. It is observed that in some grammineae as rice (*Oryza sativa*), silica constitutes 20–22% of its total production in the rice husk form [23]. Sugarcane bagasse contains around 5.08 to 7.08% of silica in dry matter basis [24], and coffee husk contains around 1 to 3% of silica in dry matter basis[25].

It is important to mention that the mechanical strength of plants resides greatly on the cell wall, enabling them to achieve and maintain erect habit conductive to light interception. There exist a relation in plants stress and the increased rigidity of cell walls of plants grown with ample available silica [26,27]. When plants die, the silica is reincorporated into the soil where microorganisms play an important role in the degradation of organic matter and in the release of minerals nutrients[28,29], other important source that raises mineralization rate is earthworms producing biohumus. Humus contains principally carbon, oxygen, hydrogen and minor proportion of other minerals. These elements vary within the humic material in order to define chemical characteristics of the original basis. There exists a symbiotic interaction between earthworms and microorganisms that breakdown and fragment the organic matter progressively, finally incorporating it into water-stable aggregates. The mineral nutrients in earthworm casts and lining earthworm burrows are in a form readily available to plants. There is evidence that interactions between earthworms and microorganisms not only provide these available nutrients, but stimulate plant growth indirectly in others ways [30]. The digestive system of earthworms consists of a pharynx, esophagus and gizzard (zone reception) followed by an anterior intestine that secrets enzymes and a posterior intestine that absorb nutrients. During progress through this digestive system, there is a dramatic increase in numbers of microorganism of up to 1,000 times. The digestive systems of earthworms from different species, genera and families differ in detail, but their gusts have a common basic structure. In different species earthworms *Eisenia foetida* is peculiar for its degradation rate [28]. Most studies of digestive enzymes in earthworms have been limited to the lumbricids. Protease, lipase, amylase, lichenase, cellulose and chitinase activities also have been described [31]. A wide range of microorganisms, including bacteria, algae, protozoa, actinomycetes, fungi and even nematodes, are found commonly throughout the length of the earthworms gut. The species of microbes in the gut are usually very similar to those in the surrounding soil or organic matter upon which the earthworms feed [32-34]. *Eisenia foetida* is considered a machine to produce humus in conditions environmental control and the microorganism can live in it in anoxic effect raising productivity in the material expel. The biological mechanism to earthworms transforms organic matter and even so carries out biosilification even is uncertain. Understanding the mechanism of silica nanofabrication in other organisms is supported by a precursor namely biosilica monosilicic acid $Si(OH)_4$ [35,36]. Proteins have been isolated from diatoms, sponges and grasses that are proposed to be responsible for biosilification and have been sequenced and some of the key amino acids identified. Other authors have studied the role of homopolymers of various amino acids that are key constituents of the proteins lysine, histidine, arginine, cysteine, proline and serine in the process biosilification [37]. This biopolymer acts as gelating agents of silica oligomers in silicic acid and as flocculation agents in silica sols [38,39]. Other researches have been focused toward the development of efficient and innovative fabrication methods to obtain inorganic materials using microorganisms from potential cheap agro-industrial waste materials and could lead to an energy-conserving and economically viable green approach toward the large-scale synthesis of oxide nanomaterials [9]. Thus, we develop a novel process for synthesis of diverse nanometric materials with specific crystal arrays as precursors to agro-industrial wastes employing annelids, an approach not used before, that permit to rise natural sources dedicated to production particles' mean biosilification.

# EXPERIMENTAL

Three sources derived from agro-industrial activity were used: rice husk, coffee husk and sugarcane bagasse. These by-products were added to vermicompost separately. The annelid specimen used was *Eisenia foetida*. The environmental conditions ideal to the reproduction and control of these specimens were set up: temperature at 20°C, moisture around 60–85%, aeration conditions and darkness. The stabilization time was around 1 month and the humus obtained was dried in a room at a temperature between 30 and 40°C. Then, the humus was sieved to size 0.5 mm approximately. Next, the sample was calcinated to eliminate the organic matter. Three temperature levels were used: 500, 600 and 700°C for each agro-industrial waste by 19 h. Calcinations were carried out in a muffle Lindberg/Eurotherm model 847 with energy consumption of 0.17 kcal/h $cm^3$, considering that the average density of the nanoparticles is around 0.1380 $g/cm^3$, the consumed energy in the calcinations is approximately 1.2318 kcal/h by each gram of recuperated $SiO_2$ particles. This energy could be considered low in comparison with other conventional process where fumed silica is manufactured with a consumed energy of until 15.48 kcal/h by each gram of $SiO_2$ particles [40]. Thus, the samples were tried with nitric and hydrochloric acids (volume ratio 3:1). For each gram of calcinated sample, 4 ml of acid mix was added in order to eliminate impurities (calcium, potassium, magnesium, manganese, iron, boron and phosphorous). Acid treatment was achieved at 40°C by 4 h with constant stirring. Then, samples were filtrated and washed with distilled water to neutralize them. Solids obtained were dried at room temperature. All reagents employed were provided by Sigma–Aldrich.

Also, as a reference, $SiO_2$ was obtained from the agro-industrial wastes without employing vermicompost bioprocess. The extraction process to recuperate $SiO_2$ is the same as described previously using calcination and acid treatment. In addition, commercial synthetic $SiO_2$ Aerosil® 130 provided by Degussa AG was employed to compare size and structure with $SiO_2$ nanoparticles produced in this research. Aerosil$_{®}$ 130 particles are amorphous $SiO_2$ nanoparticles produced by high-temperature hydrolysis of silicon tetrachloride in an oxygen gas flame [41]. Also, this research compares the particle features based on biotransformation process with those synthesized using chemical process.

$SiO_2$ powders were characterized by employing a Fourier transform infrared spectrophotometer (FTIR) Bruker Vector 33 using KBr powders. Transmission electron microscopy (TEM) was realized using a JEOL TEM-1010 transmission electron microscope, and high-resolution transmission electron microscopy (HRTEM) was realized in a Tecnai G2 T20 Microscope. Average particle size was determined for dynamic light scattering (DLS) using a Brookhaven model BI200SM with laser He–Ne of 35 mW model 9167 EB-1 Melles-Griot. Elemental analysis was carried out using energy dispersion spectroscopy (EDS) mean software Oxford Inca X–Sight. EDS is adapted in equipment JEOL JSM-6060 LV Scanning Electron Microscope. X-ray diffraction (XRD) was realized in a diffractometer Rigaku, model MiniFlex, with a wavelength from 1.54 Å corresponding kα cupper radiations. Crystalline structures present in the samples were analyzed with Materials Data Jade software of MDI Materials Data.

# RESULTS AND DISCUSSION

Figure 1 shows the FTIR spectroscopy analysis to the samples obtained from the agro-industrial wastes: rice husk, coffee husk and sugarcane bagasse after vermicompost bioprocess, calcinations and extraction process. The spectra present three important bands that allow identifying the $SiO_2$. At 1,080 $cm^{-1}$, a band corresponds to stretching antisymmetric mode of Si–O–Si group. Around 800–810 $cm^{-1}$, bending vibration mode is detected. This peak corresponds to Si–O group. Also, the peak observed at 500 cm$^{-1}$ corresponds to rocking mode of the Si–O group. At 3,500 and 1,640 cm$^{-1}$ are observed stretching vibration mode of O–H and twisting vibration mode of H–O–H, respectively. The peaks formed at different temperatures of calcination (500, 600 and 700°C) do not show important differences in the infrared analysis. The FTIR results suggest that carbon from the organic matter is removed. Thus, silica is released, and then the suboxides found after calcinations are separated correctly. In addition, the SiO$_2$ typical bands represent strong evidence of the efficiency in the synthesis process based on biotransformation with annelids. Also, the bands corresponding to O–H vibration mode and H–O–H twisting vibration mode indicate that particles synthesized remain hydrated. As can be observed in Fig. 1, commercial sample Aerosil$_®$ 130 shows the same three bands corresponding to SiO$_2$. However, the peaks at 3,500 and 1,640 cm$^{-1}$are weaker than the peaks belonging to the samples synthesized using vermicompost. The latter allows to assume that nanoparticles obtained by vermicompost are further hydrated than Aerosil®130 particles. This could be important for developing hybrid materials, inasmuch as these types of materials use silanol groups to attach organic moieties to inorganic Si [42-45].

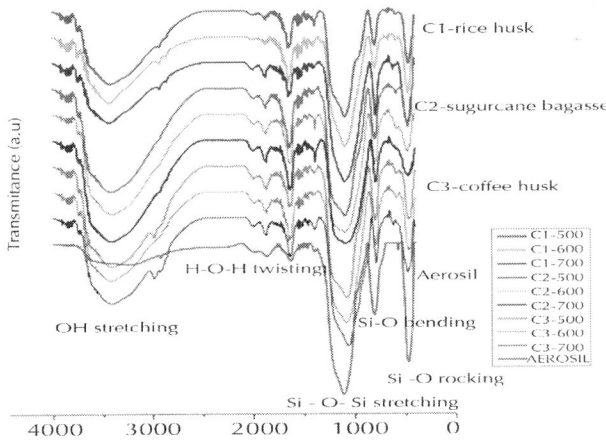

**Figure 1:** IR bands to synthesis of nanoparticles SiO$_2$ obtained for vermicompost. The Figure show the bands more representatives to SiO$_2$to in rice husk, sugarcane bagasse and coffee husk at temperature of calcination: 500, 600 and 700°C. Representative bands for Aerosil$_®$ 130 too are shown.

Figures 2a–2f show TEM images for the $SiO_2$ particles obtained using the synthesis by vermicompost. The morphology and structure of these particles is not completely spherical. In addition, several particle clusters are observed in these images. This agglomerates show different shapes of nanometric size. Figure 2g–2h shows $SiO_2$ particles obtained from the agro-industrial wastes without biotransformation. In these pictures are observed elliptical particles with different diameter and a relatively bigger size than particles analyzed in the Fig. 2a–2f. This reveals that bioprocesses employed contribute to reduce the size in $SiO_2$ particles. We suggest that this size reduction is induced in earthworm gut with the contribution from several microorganisms. Figure 3a–3b shows HRTEM images of $SiO_2$ particles obtained through bioprocess. In these images, it is possible appreciate the $SiO_2$ particles in an atomic scale. Nanometric size and crystal structure are observed in these images. Thus, it is shown that these particles present as well a crystalline structure contrary to amorphous synthetic $SiO_2$. In addition, the Fig. 3 shows the diffraction patterns and Miller index corresponding with these structures in a selected area (A). Figures show the nanoparticles obtained from the precursors that contain a higher content of $SiO_2$ such as rice husk (Fig. 3a) and sugarcane bagasse (Fig. 3b). The indexation of nanoparticles obtained from rice husk shows $SiO_2$ with crystal arrangement related to α hexagonal quartz with the next lattice parameters: 0.340 nm (1−1 0 0), 0.243 nm (1 1−2 0) and 0.198 nm (2 0−2 0). In the case of nanoparticles obtained from sugarcane bagasse, $SiO_2$ possess a tetragonal arrangement with the next lattice parameters: 0.200 nm (2 0 0), 0.199 nm (0 2 0) and 0.142 nm (2 2 0). In accord with these results, it is possible to suppose an important contribution of the vermicompost bioprocess in order to give crystalline phases in $SiO_2$ nanoparticles.

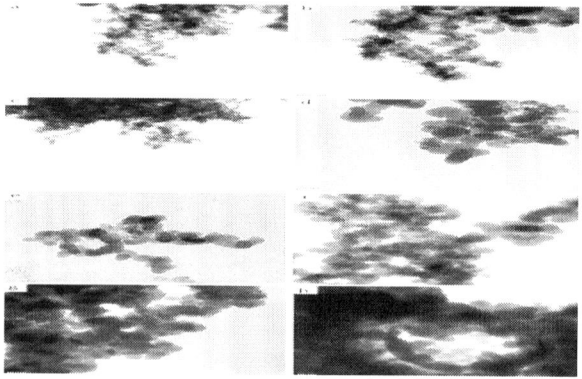

**Figure 2:** TEM images to synthesis of $SiO_2$ obtained with bioprocess (a–f) and without bioprocess (g–h). Figure 2 shows differences in size, form and

dispersity to nanoparticles synthesized for both methods employed, with bioprocess (a–f) and without bioprocess (g–h).

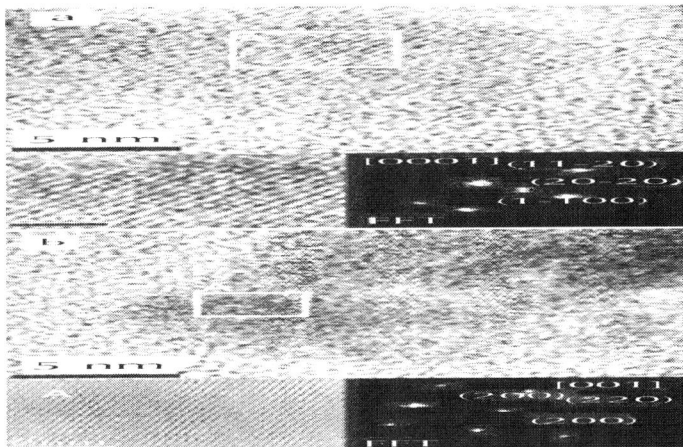

**Figure 3:** HRTEM images and indexation of SiO$_2$ nanoparticles obtained by bioprocess: a nanoparticles obtained from rice husk b nanoparticles obtained from sugarcane bagasse.

In order to confirm the TEM and HRTEM results, particle size was analyzed employing dynamic light scattering (DLS). The results obtained by this technique are summarized in Table 1. The mean diameter of particles synthesized employing bioprocess is around 81 nm and do not show polydispersity. In contrast, the mean diameter of particles synthesized without using bioprocess is between 152 and 254 nm with low polydispersity. This is in agreement with TEM and HRTEM images and confirms the contribution of bioprocess in particle size reduction to achieve a nanometric range. It is important to emphasize the presence of agglomerates in the particles obtained without biotransformation, which are not observed in the particles obtained by employing bioprocess. Thus, as well as the bioprocess contribution to particle size reduction also provides specific arrangement to the particles. It is suggested that this effect is produced by the microorganisms which are part of metabolism in the earthworm. The size and morphology of particle in the biosilification process are related with the concentration of inorganic phosphate and polyamines, both compounds play an important role in order to catalyze the polycondensation of silanol groups [46]. In vermicompost, the concentrations of these compounds vary with the microbial population. Thus, it is supposed that the size and morphology of $SiO_2$ nanoparticles could be changed in this kind of bioprocess.

**Table 1:** Diameter mean using DLS to particles $SiO_2$ synthesized with bioprocess and without bioprocess

| Sample | With bioprocess | | Without bioprocess | |
|---|---|---|---|---|
| | Diameter mean (nm) | Variance relative | Diameter mean (nm) | Variance relative |
| Rice husk | 81.1 | 0.018 | 233.4 | 0.025 |
| Sugarcane bagasse | 81.1 | 0.041 | 185.1 | 0.000 |
| Coffee husk | 81.8 | 0.014 | 152.6 | 0.036 |
| Aerosil® 130 | – | – | 75.8 | 0.000 |

Table 1 shows variations in mean diameter in nm and variance relative to particles $SiO_2$ synthesized since precursors: rice husk, sugarcane bagasse and coffee husk with bioprocess and without bioprocess. A comparative with Aerosil® 130 too is presented

Espíndola-Gonzalez *et al.*

Espíndola-Gonzalez *et al. Nanoscale Research Letters* 2010 **5**:1408-1417, doi:10.1007/s11671-010-9654-6

Table 2 shows the elemental analysis for $SiO_2$ particles synthesized by bioprocess; a significant amount of silicon and oxygen weight percent is observed in all samples obtained with the precursors employed: 41.62% Si and 52.90% O for particles obtained from rice husk; 23.37% Si and 42.37% O in particles synthesized from sugarcane bagasse and 41.68% Si and 55.39% O to particles extracted from coffee husk. Also, in the composition are observed other elements with lower percent than silica and oxygen, such as sodium, magnesium, aluminum, potassium and calcium. These elements are typical in the biomineralization process and take part in the crystalline phases growing that are produced with this biological mechanism.

**Table 2:** EDS to particles SiO$_2$ synthesized with bioprocess and without bioprocess

| Rice husk | | | Sugarcane bagasse | | | Coffee husk | | |
|---|---|---|---|---|---|---|---|---|
| Element | % weight | % Atomic | Element | % weight | % Atomic | Element | % weight | % Atomic |
| O K | 52.90 | 66.82 | C K | 31.35 | 42.02 | O K | 55.39 | 68.88 |
| Na K | 0.350 | 0.310 | O K | 42.37 | 42.91 | Na K | 0.540 | 0.470 |
| Mg K | 0.790 | 0.650 | Na K | 0.410 | 0.290 | Al K | 0.510 | 0.380 |
| Al K | 1.030 | 0.770 | Al K | 1.140 | 0.690 | Si K | 41.68 | 29.52 |
| Si K | 41.62 | 29.95 | Si K | 23.37 | 13.48 | K K | 0.900 | 0.460 |
| K K | 1.390 | 0.720 | K K | 0.680 | 0.280 | Ca K | 0.330 | 0.160 |
| Ca K | 0.450 | 0.250 | Ca K | 0.610 | 0.250 | Mo L | 0.650 | 0.130 |
| Ti K | 0.350 | 0.150 | Fe K | 0.280 | 0.080 | | | |
| Fe K | 1.120 | 0.410 | | | | | | |
| Total | 100.0 | | Total | 100.0 | | Total | 100.0 | |

Table 2 shows EDS analysis in percent weight and percent atomic to nanoparticles SiO$_2$ employing bioprocess in rice husk, sugarcane bagasse and coffee husk

Espíndola-Gonzalez et al.

Espíndola-Gonzalez et al. *Nanoscale Research Letters* 2010 **5**:1408-1417, doi: 10.1007/s11671-010-9654-6

Figure 4 shows the diffractrograms generated from the particles obtained using the three precursors: rice husk (a–b), sugar cane bagasse (c–d) and coffee husk (e–f). Crystalline nanoparticles are obtained using two different calcination temperatures: 500°C (a, c and e) and 700°C (b, d and f). Figure 4a shows the diffractrogram corresponding to particles obtained from rice husk at 500°C. In this figure, it is observed an amorphous structure when the bioprocess is not used during the synthesis. In contrast, crystalline phases are found when the bioprocess is employed. In this diffractrogram, it is possible to identify typical peaks corresponding to diffraction planes from quartz and other polymorphic structures of $SiO_2$, such as trydimite. Also, erionite and albite diffraction planes corresponding to aluminum silicate groups are found. These results support the proposal that biomineralization mechanism is achieved through earthworms contributing in the atomic arrangement and modifying the original structure toward specific crystalline structures. Figure 4b shows the diffractrogram corresponding to particles obtained from rice husk at 700°C. In spite of the high temperature, even amorphous structure is observed when bioprocess is not used and only one diffraction plane corresponding at low quartz is shown. This crystal is formed probably by temperature effect. In the same Fig. (4b), it is shown the diffractrogram corresponding to the samples treated with bioprocess at 700°C. In this figure, the crystalline structure is related to low quartz, trydimite and albite.

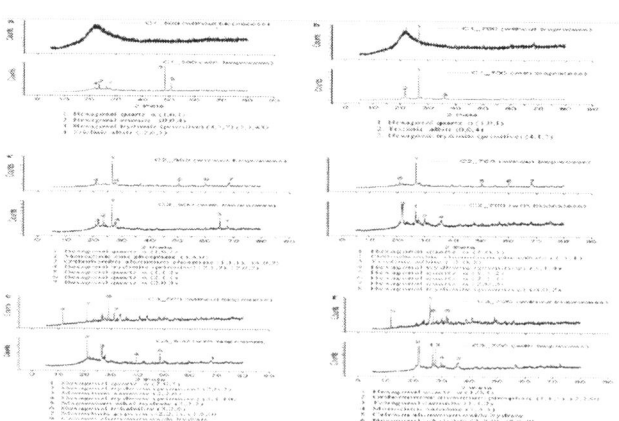

**Figure 4:** X-ray diffraction to synthesis of nanoparticles $SiO_2$ obtained with bioprocess and without bioprocess using as precursors: rice husk (a–b) sugar-

cane bagasse (c–d) and coffee husk (e–f) to two temperatures of calcinations: 500 and 700°C. It shows a change in the crystallinity of the nanoparticles synthesized, since the different precursors used and varying the temperature of calcination and employing or not employing the bioprocess.

Figure 4c and 4d show the diffractrograms corresponding to particles obtained from sugarcane bagasse. Crystalline phases are present in both procedures (with and without bioprocess). However, some diffraction planes are not the same in both processes. This allows to assume that the calcination temperature and earthworm metabolism play an important role to generate different crystalline phases that involve not only silicon and oxygen atoms, but other elements take part, as well. In Fig. 4c and 4d, by comparing the diffractrograms (samples obtained without bioprocess at 500 and 700°C), it is possible to observe that the same crystalline phases are found: quartz (different crystallographic planes) and trydimite. However, the diffractrograms corresponding to samples obtained via bioprocess show different crystalline phases, such as: zinc phosphate, aluminum phosphate at 500°C and albite in 700°C.

It is important to mention that silica shows several polyphorms depending on temperature and pressure. Thus, although bioprocess conditions employed to obtain $SiO_2$ particles are favorable to inducing α hexagonal quartz (corroborated by X-ray diffraction), it is possible to find others metastable polyphorms such as tetragonal arrangement (268–1,470°C). This structure belongs to β crystobalite [3]. Therefore, some $SiO_2$ nanoparticles obtained from sugarcane bagasse can be found with transitions of hexagonal to tetragonal phase. Thus, tetragonal arrangement does not appear in X-ray diffraction; however, in some nanoparticles characterized by HRTEM (Fig. 3b) it is identified.

Diffractrograms corresponding to particles obtained from coffee husk are shown in Fig. 4e and 4f. Significant changes in both processes (using bioprocess and without employing bioprocess), which are produced by temperature and metabolism in earthworms, are observed. At 500°C, by employing bioprocess, the diffraction peaks detected are related to quartz, trydimite, sanidine and magnesium nickel hydride, meanwhile without bioprocesses diffraction peaks appear, corresponding to trydimite, gypsum and calcium aluminim oxide hydrate. At 700°C, by employing bioprocess, quartz, aluminim phosphate and caminite are found, meanwhile without bioprocess appears: calcium aluminim oxide hydrate, trikalsilite and quartz.

Table 3 shows the crystallization percent, considering precursor type and calcination temperature employed. Crystallization degree is obtained for both: particles obtained via bioprocess and particles synthesized without using bioprocess. These values are calculated by considering mean low curve area in the peaks from XRD. It is observed that, for rice husk, the effect in the temperature is not enough as to transform the amorphous phase. However, the metabolic effect in earthworms increases the crystalline phase up to 28.47% (500°C) and 16.69%(700°C). In sugarcane bagasse, the effect is the opposite, the crystalline phase is low for the samples obtained via bioprocess and crystallinity percent increases for particles obtained without bioprocess. In addition, diffractrograms obtained from sugarcane bagasse show that some crystalline phases do not correspond to $SiO_2$, such as zinc phosphate and aluminim phosphate; however, when the bioprocess is not used, particles show a considerable number of $SiO_2$ polymorphs. This suggests that most of the oxygen needed to form $SiO_2$ may be consumed by the earthworms during their metabolic process. This assumption is in agreement with EDS analysis, inasmuch as EDS results for particles obtained from sugar baggasse show lower percent of oxygen than particles synthesized from rice husk and coffee husk. With respect to the crystallization percent in the particles obtained from coffee husk, there is not a clear tendency. For the samples obtained using bioprocess at 500°C, the crystallization percent is lower than that of samples synthesized without bioprocess. Probably, bioprocessing contributes to provide certain array in the nanoparticles, depending on organic matter, oxygen and other inorganic compounds present in the precursors. This would confirm that temperature is not the most important factor for the crystallinity in these kinds of nanoparticles.

**Table 3:** Percent crystallization in particles $SiO_2$ synthesized with bioprocess and without bioprocess

| Sample | Crystallization (%) | | | |
|--------|----------------|--------|--------|--------|
| | With bioprocess | | Without bioprocess | |
| | 500°C | 700°C | 500°C | 700°C |
| Rice husk | 28.47 | 16.69 | 0.000 | 1.910 |
| Sugarcane bagasse | 10.42 | 9.190 | 76.12 | 13.71 |
| Coffee husk | 15.73 | 25.82 | 58.28 | 17.36 |

In this table is presented an estimative obtained for low curve areas in the peaks of XRD. These percents shows the contribution than the temperature of calcination and the bioprocess have on the crystallization of nanoparticles of $SiO_2$

Espíndola-Gonzalez *et al.*

Espíndola-Gonzalez *et al. Nanoscale Research Letters* 2010 **5**:1408-1417, doi:10.1007/s11671-010-9654-6

# CONCLUSIONS

The microbial population in annelids is very important to achieve the biotransformation of the amorphous silica naturally present in the analyzed agro-industrial wastes. Characteristics of the organic matter exposed to a broad variety of microorganisms, as well as the method employed for the fragmentation and minerals release, represent key factors to understand the biocrystallization. Our results reveal a novel synthesis method to obtain $SiO_2$ crystalline nanoparticles using annelids' biotransformation by employing agro-industrial wastes. The approach represents an inexpensive and relatively eco-friendly technology in comparison with standard chemical methods. By taking into account the biological aspects of the production of $SiO_2$ nanoparticles with specific crystal arrangement using annelids allows to extend the number of living organism dedicate to biosilification, in addition to open an interesting field toward the knowledge of new annelid bioprocesses with a primitive metabolism, as potential alternative natural nanotechnology bioprocesses to synthesize nanoparticles and nanostructures, for the particles obtained through biotransformation by annelids show similar characteristics than synthetic $SiO_2$ Aerosil®130 such as size, composition and polydispersity. While synthetic particles possess amorphous structure the particles synthesized using vermicompost present different disables characteristics such as crystalline (different polymorphism) and nanometric dimension.

# ACKNOWLEDGMENTS

The authors are grateful to Ms. Maria de Lourdes Palma for her assistance in TEM, to Dr. Genoveva Hernández-Padron for her assistance in IR analysis, to Dr. Eric Rivera for his assistance in XRD, to CINVESTAV Querétaro, particularly to Dr. S. Jimenez and Mr. F Rodriguez for their assistance in some measurements and to DGEST and Consejo Nacional de Ciencia y Tecnologia (CONACyT), Mexico, for the economic support through the projects P333-05 and JI-58232, respectively. Financial support from the National Council for Science and Technology of Mexico (CONACYT) (PhD Scholarship to A.E.-G.) is gratefully acknowledged.

# REFERENCES

1. Alexandre A, Meunier J-D, Colin F, Koud J-M: Plant impact on the biogeochemical cycle of silicon and related weathering processes. *Geochim. Cosmochim. Acta* 1997, 61:677-682. COI number [1:CAS:528:DyaK2sXhsFaltbw%3D]; Bibcode number [1997GeCoA..61..677A].

2. Morse ED: Silicon biotechnology: harnessing biological silica production to construct new materials. *Els. Sci. Trends Biotech.* 1999, 17:230-232. COI number [1:CAS:528:DyaK1MXltFaisL8%3D]

3. Iler RK: *The Chemistry of Silica.* John Wiley & Sons, New York; 1979.

4. Ball P: *Made to Measure: New Materials for the 21st Century.* Princenton University Press, Princenton, NJ. USA; 1999.

5. Bauerlein E: Biomineralization of unicellular organism: an unusual membrane biochemistry for the production of inorganic

nano- and microstructures. *Angew Chem. Int. Edn* 2003, 42:614-641. COI number [1:CAS:528:DC%2BD3sXhs1Sitbw%3D]

6.  Hamm CE, *et al.*: Architecture and material properties of diatoms shells provide effective mechanical protection. *Nature* 2003, 421:841-843. COI number [1:CAS:528:DC%2BD3sXht1Kqt7Y%3D]; Bibcode number [2003Natur.421..841H]

7.  Asada R, Okuno M, Tazaki K: Structural anisotropy of biogenic silica in pennate diatoms under Fourier transform polarized infrared spectroscopy. *J. Mineral Petrol. Sci.* 2002, 97:219-226. COI number [1:CAS:528:DC%2BD3sXlsVOitQ%3D%3D].

8.  Almqvist N, *et al.*: Micromechanical and structural properties of pennate diatom investigated by atomic force microscopy. *J. Microsc.* 2001, 202:518-532. COI number [1:CAS:528:DC%2BD3MXlt1SrsLc%3D].

9.  Bansal V, Ahmad A, Sastry M: Fungus-mediated biotransformation of amorphous silica in rice husk to nanocrystalline silica. *J. Am. Chem. Soc.* 2006, 128:14059-14066. COI number [1:CAS:528:DC%2BD28XhtVKitbzN].

10. Inagaki F, Motomura Y, Ogata S: Microbial silica deposition in geothermal hot waters. *Appl. Microbiol. Biotechnol.* 2003, 60:605-611. COI number [1:CAS:528:DC%2BD3sXitlOqsbg%3D].

11. Kastner M: *The Oceanic Lithosphere*. Wiley, New York; 1981.

12. Dujardin E, Mann S: Bio-inspired materials chemistry. *Adv. Mater.* 2002, 14:775. COI number [1:CAS:528:DC%2BD38XkvVOiurw%3D].

13. Vrieling EG, Beelen TPM, van Santen RA, Gieskes WWC: Nanoscale uniformity of pores in diatomaceous silica: a combined small and wide angle X-ray scattering study. *J. Phycol.* 2003, 35:1044-1053.

14. Vrieling EG, Beelen TPM, van Santen RA, Gieskes WWC: Diatoms silicon biomineralization as an inspirational source of new approaches to silica production. *J. Biotechnol.* 1999, 70:41-53.

15. Sun Q, Vrieling EG, van Santen RA, Sommerdijk NAJM: Bioinspired synthesis of mesoporous silicas. *Solid State Mater. Sci.* 2004, 8:111-120. COI number [1:CAS:528:DC%2BD2cXlvVWru70%3D].

16.  Dugdale RC, Wilkerson FP: Silicate regulation of new production in the equatorial Pacific upwelling. *Nature* 1998, 391:270-273. COI number [1:CAS:528:DyaK1cXnsVyruw%3D%3D]; Bibcode number [1998Natur.391..270D].

17.  Hecky RE, Mopper K, Kilham P, Degens ET: The amino acid and sugar composition of diatom cell-walls. *Marine Biol.* 1973, 19:323. COI number [1:CAS:528:DyaE3sXlt1art7w%3D].

18.  Kroger N, Lorenz S, Brunner E, Sumper M: Self-assembly of highly phosphorylated silaffins and their function in biosilica morphogenesis. *Science* 2002, 298:548.

19.  Shimizu K, Cha J, Stucky GD, Morse DE: Silicatein $\alpha$: cathepsin L-like protein in sponge biosilica. *Proc. Natl. Acad. Sci.* 1998, 95:6234. COI number [1:STN:280:DyaK1c3mtleguw%3D%3D]; Bibcode number [1998PNAS...95.6234S].

20.  Perry CC, Keeling-Tucker T: Model studies of colloidal silica precipitation using biosilica extracts from Equisetum telmateia. *Colloid Polym. Sci.* 2003, 281:652. COI number [1:CAS:528:DC%2BD3sXkvFent74%3D].

21.  Derry AL, Kurtz CA, Ziegler K, Chadwick AO: Biological control of terrestrial silica cycling and export fluxes to watersheds. *Nature* 2005, 433:728-730. COI number [1:CAS:528:DC%2BD2MXhtle qsLo%3D]; Bibcode number [2005Natur.433..728D].

22.  Eipstein E: The anomaly of silicon in plant biology. *Proc. Natl. Acad. Sci. USA* 1994, 91:11-17. Bibcode number [1994PNAS...91...11E].

23.  Ding TP, Ma GR, Shui MX, Wan DF, Li RH: Silicon isotope study on rice plants from the Zhejiang province. *China. Chin. Chem. Geol* 2005, 218:41. COI number [1:CAS:528:DC%2BD2MXktV ahsLw%3D].

24.  Suryawanshi BG, Patil SS, Patil BN: Studies on the chemical composition of sugarcane tops. *J. Maharashtra Agric. Univ.* 2003, 28:50-51.

25.  Pandey A, *et al.*: Biotechnological potential of coffee pulp and coffee husk for bioprocesses. *J. Biochem. Eng.* 2000, 6:153-162. COI number [1:CAS:528:DC%2BD3cXntFShsLo%3D].

26.  Raven JA: The transport and function of Silicon in plants. *Biol. Rev.* 1983, 58:179-207. COI number [1:CAS:528:DyaL3sXlslakt7s%3D].

27. Jones LHP, Handreck KA: Silica in soils, plants and animals. *Adv. Agron.* 1967, 19:107-149. COI number [1:CAS:528:DyaF1cXitFWrsA%3D%3D]

28. Edwards CA, Lofty Jr: *Biology Earthworms*. Chapman and Hall, London; 1977.

29. Lee KE: *Earthworms. Their Ecology and Relationships with Soils and Land Use*. Academic Press, Australia; 1985.

30. Edwards AC, Fletcher EK: Interactions between earthworms and microorganisms in organic-matter breakdown. *Agric. Ecosystems Env.* 1988, 24:235-247.

31. Laverack : The physiology of earthworms. *Int. Ser. Monogr. Pure Appl. Biol., Zool* 1963, 15:206.

32. Went JC: Influence if earthworms on the number of bacteria in the soil. In *Soil Organisms*. Edited by Koeksen J, Drift J. North Holland Publishing Company, Amsterdam; 1963.

33. Parle JN: Microorganisms in the intestines of earthworms. *J. Gen. MIcrobiol.* 1963, 31:1-11.

34. Atlavinyte O, Daciulyte J, Lugauskas A: Correlation between the numbers earthworms microorganisms and vitamin B12 in soils fertilized with straw. *Liet. TSRA Mokslu Akad. Darb. Ser. B* 1971, 3:43-56.

35. Treguer P, *et al.*: The silica balance in the World Ocean: a reestimate. *Science* 1995, 268:375-379. COI number [1:CAS:528:DyaK2MXltFKkt7k%3D]; Bibcode number [1995Sci...268..375T].

36. Del Amo YB: The chemical form of dissolved Si taken up by marine diatoms. *A.M. J. Phycol.* 1999, 35:1162-1170. COI number [1:CAS:528:DC%2BD3cXnsFCqsA%3D%3D].

37. Patwardhan VS, Clarson JS: Silicification and biosilicification part 6. Poly-l- histidine mediated synthesis of silica at neutral pH. *J. Inorg. Organomet. Poly.* 2003, 13(1):50-53.

38. Coradin T, Durupthy O, Livage J: Interaction of amino-containing peptides with sodium silicate and colloidal silica: a biomimetic approach of silification. *Langmuir* 2002, 18:2331-2336. COI number [1:CAS:528:DC%2BD38Xhtlyjt7o%3D].

39. Coradin T, Roux C, Livage J: Biomimetic self-activated formation of multiscale porous silica in the presence of arginine-based

surfactants. *J. Mater. Chem.* 2002, 12(5):1242-1244. COI number [1:CAS:528:DC%2BD38XivF2rtrk%3D].

40. Baluais G, Caratini Y: Medium purity metallurgical silicon and method for preparing same. *Patent US7404941 assigned to Ferropem* 2005.

41. *Aerosil R-Manufacture, properties and applications, Technical Bullettin Pigments.* N. 11, Degussa-Huls AG, Germany; 2002.

42. Sales AAJ, Petrucelli CG, Oliveira EVJF, Airoldi C: Some features associated with organosilane groups grafted by the sol–gel process onto synthetic talc-like phyllosilicate. *J. Collod. Inter. Sci.* 2006, 297:95-103. COI number [1:CAS:528:DC%2BD28Xisl2qt7o%3D].

43. Lim HM, Blanford FC, Stein A: Synthesis and characterization of a reactive vinyl-functionalized mcm-41: probing the internal pore structure by a Bromination reaction. *J. Am. Chem. Soc.* 1997, 119:4090-4091. COI number [1:CAS:528:DyaK2sXisVyjsbk%3D]
.

44. Melde JB, Johnson JB, Charles TP: Mesoporous silicate materials in Sensing. *Sensors* 2008, 8:5202-5228. COI number [1:CAS:528:DC%2BD1cXhsVekt7zK].

45. Gao X, Jensen ER, Li W, Deitzel J, Mcknight H, Gillespie WJ Jr.: Effect of fiber surface texture created from silane blends on the strength and energy absorption of the glass fiber/epoxy interphase. *J. Compos. Mater* 2008, 42:513. COI number [1:CAS:528:DC%2BD1cXltFyns70%3D].

46. Sumper M, Kroger N: Silica formation in diatoms: the function of long-chain polyamines and silaffins. *J. Mater. Chem.* 2004, 14:2059-2065. COI number [1:CAS:528:DC%2BD2cXls1egsrs%3D].

# Citations

# CHAPTER 1

Christian Löser, Thanet Urit, Erik Gruner, and Thomas Bley , "Efficient growth of Kluyveromyces marxianus biomass used as a biocatalyst in the sustainable production of ethyl acetate ", doi:10.1186/s13705-014-0028-2.

# CHAPTER 2

Bárbara G Paes and João RM Almeida, Genetic Improvement of Microorganisms for Applications in Biorefineries, doi:10.1186/s40538-014-0021-1.

# CHAPTER 3

Claudia Elizabeth Thompson, Walter Orlando Beys-da-Silva, Lucélia Santi, Markus Berger, Marilene Henning Vainstein, Jorge Almeida Guima rães, and Ana Tereza Ribeiro Vasconcelos, A potential source for cellulolytic enzyme discovery and environmental aspects revealed through metagenomics of Brazilian mangroves, doi:10.1186/2191-0855-3-65.

# CHAPTER 4

Ramesh P Babu, Kevin O'Connor, and Ramakrishna Seeram, Current Progress on Bio-Based Polymers and Their Future Trends, doi:10.1186/2194-0517-2-8.

# CHAPTER 5

Latika Bhatia, Sonia Johri, and Rumana Ahmad, An economic and ecological perspective of ethanol production from renewable agro waste: a review, doi: 10.1186/2191-0855-2-65.

# CHAPTER 6

Jinlong Guo, Liping Xu, Yachun Su, et al., "ScMT2-1-3, a Metallothionein Gene of Sugarcane, Plays an Important Role in the Regulation of Heavy Metal Tolerance/Accumulation," BioMed Research International, vol. 2013, Article ID 904769, 12 pages, 2013. doi:10.1155/2013/904769.

# CHAPTER 7

Rosa Estela Quiroz-Castañeda and Jorge Luis Folch-Mallol (2013). Hydrolysis of Biomass Mediated by Cellulases for the Production of Sugars, Sustainable Degradation of Lignocellulosic Biomass - Techniques, Applications and Commercialization, Dr. Anuj Chandel (Ed.), ISBN: 978-953-51-1119-1, InTech, DOI: 10.5772/53719.

# CHAPTER 8

A Espíndola-Gonzalez, AL Martínez-Hernández, C Angeles-Chávez, VM Castaño, and C Velasco-Santos, novel crystalline sio2 nanoparticles via annelids bioprocessing of agro-industrial wastes, doi:10.1007/ s11671-010-9654-6.

# Index